GOD'S KEYS TO UNLOCKING HEALTH, HEALING, AND WHOLENESS

FREQUENCIES OF HEAVEN

GOD'S KEYS TO UNLOCKING HEALTH, HEALING, AND WHOLENESS

FREQUENCIES OF HEAVEN

Dr. Scott Oatsvall,
Joan Hunter,
with John Tussey

Published by Hunter Books, an imprint of Joan Hunter Ministries
PO Box 411, Pinehurst, TX 77362 USA | JoanHunter.org, JoanHunter.ca
ISBN 978-0-9988739-7-8 (print); 978-0-9988739-8-5 (ebook)

Cover by Yvonne Parks/Pear Creative
Editing/interior by Avodah Editorial Services

Printed in the United States of America
25 26 27 28 29 30 5 4 3

CONTENTS

INTRODUCTION

What if the history we've been taught is not the truth? What if there are hidden, suppressed technologies—ancient and advanced—that could change everything we know about the world? What if I told you we are caught in the midst of a secret war, a war not just of physical power, but of frequencies? This is a battle that spans both the material and spiritual realms, a covert conflict that few are even aware of. Behind the scenes, powerful forces are actively working to keep these frequencies and technologies from being revealed to the public.

But why the secrecy?

For centuries, a small group of elite has held tight to knowledge that could radically alter our world. What would happen to the global powers—the titans of industry, the religious authorities, the political rulers—if the truth came to light? What if we learned that the pyramids were built using fallen angel technology? Or that there are frequencies capable of curing all disease, including cancer?

The answer is simple: power, not money.

There are energy sources—technologies that have been known for decades—that could free humanity from many of its greatest struggles. Nikola Tesla, for example, developed systems for energy and propulsion that were smaller than a modern laptop, capable of running entire households or cars, powered by the electromagnetic field that surrounds us. This energy is self-sustaining, constantly replenishing itself.

But these technologies remain suppressed.

Instead, we continue to rely on outdated, harmful methods of energy production: fossil fuels, coal, nuclear power. Why? Because they're centralized. They are controlled by a select few, who hold the reins of global governance and geopolitics. Elites control the frequencies—and with them, the power to maintain their dominance.

This isn't about money in the traditional sense. This is about power—power to shape the future, power to manipulate the flow of energy, and ultimately, the power to control humanity. This is part of a much larger agenda—one that aims not only to corrupt and defile the human genome, but also to destroy the very planet that sustains us.

Frequencies of Heaven is a journey into the hidden realms of energy, technology, and consciousness. A journey that will challenge everything you thought you knew about the world, and reveal the secret war being waged for control of the frequencies that govern our lives. But there is a frequency that has the power and potential to set you free! This is the frequency of heaven.

Committed to your health,

Dr. Scott Oatsvall

CHAPTER 1
PERFECT HARMONY AND BALANCE

JOAN HUNTER

My parents, Charles and Frances Hunter, taught about the effects of frequencies on a person's health in their book *How to Heal the Sick,* which was published in January 1983. I also taught on the relationship between health and frequencies in my book *Healing the Whole Man Handbook*, which was published in July 2006, but I have been teaching this subject at my healing seminars since the year 2000.

Chapter 12 in my book *Healing the Whole Man Handbook* is called "The Immune System, Electrical and Magnetic Frequencies, Prions and the pH Balance." I teach that when all these elements in our life are active and balanced, we will have a healthy immune system that can fight off any kind of sickness, disease, or virus. This is key to getting healthy and staying well.

I also have a prayer for use daily to maintain a proper balance of our electrical, chemical, and magnetic frequencies. Clearly, our health is dependent on our immune system's ability to thwart any kind of sickness or disease.

When Dr. Scott Oatsvall scanned me early in 2024, he

said, "In all my years I have never seen anybody with such a strong immune system!" Hallelujah! That's why I can travel around the world and not get sick. I have a very strong immune system. Traumatic events happen in all our lives, and we all deal with stress. Stress happens, and when we go through stressful events, our bodies produce stress hormones that can weaken or destroy the immune system. The spirit of fear is part of that stress. Fear attacks our immune system and reduces our strength.

Our bodies are composed of a mass of water and tissues that are full of chemical, magnetic, and electrical impulses. Our spirit is constantly interacting with these electromagnetic and chemical impulses in our bodies by our thoughts, emotions, actions, and words. Our spiritual life is also affected by our physical health. As is stated in Psalm 139:14, "I am fearfully and wonderfully made." What affects one part of a person will affect the entire human being. It appears that modern scientific thinking is only now beginning to understand the true nature of the interactions of the chemical, electrical, and magnetic frequencies in human beings.

Since our bodies and our health are affected by many different factors, it is important that we understand how to deal with them properly. In this chapter I will present some of the more important ways our bodies are affected by these frequencies and what we need to do to protect and maintain our health.

Since these factors are not normally given a great deal of emphasis in medical or Christian circles, I believe this will be of great help to you. Every person must choose to take full responsibility for his or her own health and not blindly follow or adhere to whatever current medical/medicinal trends in our era. The creation, marketing, and implementation of many of these medical practices and pharmaceutical products are often driven by the combination of profit motive and insurance company policies rather than the desire to heal the sick.

During the COVID-19 pandemic, I wrote *Annihilate Fear* because I felt the heart of God crying over how the body of Christ was living in fear when we should not be. Fear is the opposite of faith and has a tremendous effect on every part of our lives. The Bible tells us, "Perfect love drives out fear" (1 John 4:18). Faith and laughter destroy stress hormones.

What happens when we laugh? Our body releases endorphins that make us feel lighter and happier while killing all the stress hormones. I'm always striving to build up my immune system, which is one reason I love to laugh. I like watching videos of little babies laughing so hard they fall over backwards. Laugh all you can. By focusing our attention on the Lord and all his many benefits rather than on our circumstances and the world, we give ourselves the freedom to enjoy the good things in life and the joy of the Lord. Our health is affected by our attitudes and our thoughts.

Electromagnetic frequencies were in perfect harmony in the garden of Eden. Sin, as well as earthquakes, has gotten earth's magnetic frequencies out of kilter. So, as we're walking around and just living, the electrical and magnetic frequencies in our body are also off kilter. It's so important that we do whatever we can do to get our own frequency back into alignment.

One of the things we can do to balance our internal magnetic/electrical frequencies is to practice grounding. Grounding is not a New Age concept. I was raised in Florida, where wearing shoes is often optional. I went everywhere without shoes in Florida. We all need to go outside more often in our bare feet. When we regularly stand on grass for fifteen minutes or more, our body's electrical field harmonizes with earth's electrical field. I recommend doing this every day if possible.

Stand outside for at least fifteen minutes every day and get some natural vitamin D in the sun. Lie down on the grass on a towel for fifteen minutes—or longer if you want. As we do this, the electrical and magnetic frequencies of the earth go into our body and reset it for healing. The frequencies of the earth are made to heal and strengthen our body.

One of the things that can lead to sickness is prions. Prions are bad cells that eat good cells, and they are designed to kill us. Some doctors have said that prions are in our food—that companies are injecting prions into our food, as well as many other products. There are those who say our food has both prions and poisons in it. We can command the prions in our body to die and to be flushed out so that they no longer affect our health.

Genetically modified organisms may have human biological products introduced into our fruits and vegetables to enhance their appearance. We need to make sure our fruits and vegetables are extremely clean of any form of parasites too. When your child or your grandchild plays in the sandbox, there are parasites in the sand. Kids are coming home from elementary school full of parasites.

Then you know what they do? They share the parasites with Mom and Dad. You are getting the tools to set yourself free as well as your family, your neighborhood, where you shop, where you work, and other places. We need to really pray over our food, because some GMOs have generational curses attached to them.

Multiple sclerosis, COVID-19 and other viruses, cancer, Epstein-Barr, and lupus all have prions. Now, is this next statement true or false? When a person is infected with a virus, then they have antibodies to help protect them from getting it again. This is false. What actually happens is that the prions stay in the body and propagate further illnesses. Additionally, when we get tired, our immune system goes down, or if we experience some kind of a trauma or stress, that opens the door for sickness to happen again.

Finally, the last of the four is our pH balance. It's very important that we maintain a proper pH balance. I always drink water that is purified and pH balanced. I travel a lot, and I run a ministry, so I have a lot of stress in my life. I handle many issues with our ministry headquarters building and our personnel, and while I'm not a building manager, I make major decisions about the building and the property.

Do you take a multivitamin every day? Many people do. Like a multivitamin, the prayer below is an everyday kind of a prayer. But the point is we have to keep our bodies clean of any and all forms of any kind of sickness, disease, virus, and prions.

I declare this prayer over you in Jesus's name:

> Father, in the name of Jesus, I come against and I curse every bit of trauma and stress in the life of

this reader and the effect that it has had on their body. A lot of people are having heart issues right now—physical heart issues—because of stress and trauma and long-term COVID-19 and other things. I thank you for healing their hearts and healing their bodies. I command their immune systems to be healed, whole, and strong—able to fight off any kind of sickness or disease. I command the electrical and magnetic frequencies in their body to return to perfect harmony and balance, in Jesus's name. I curse every prion in their body and command them to die in Jesus's name. Some of those prions have come into their body over the last couple of years and some have been there since birth. Either way, every prion is cursed in their body in the name of Jesus. I also command their pH to be completely restored to normal in Jesus's name. Any kind of heartburn, acid reflux, or other illness has to go in Jesus's name. Amen.

And now you pray this for yourself:

Father, in the name of Jesus, I come against and curse every bit of trauma and stress in my life and the effect that it has had on my body. I thank you for healing my heart and body. I command my immune systems to be healed, whole, and strong—able to fight off any kind of sickness or disease. I command the electrical and magnetic frequencies in my body to return to perfect harmony and balance, in Jesus's name. I curse every prion in my body and command

> them to die in Jesus's name. I also command my pH to be completely restored to normal in Jesus's name. Any kind of heartburn, acid reflux, or other illness must go in Jesus's name. Amen.

As you read the rest of this book, I encourage you to continue to stand in faith to unlock the fullness of health, healing, and wholeness in your life.

CHAPTER 2
UNDERSTANDING FREQUENCIES

SCOTT OATSVALL

God has created everything with a unique frequency. From the hum of a distant star to the vibrations of a leaf in the wind, frequency is a fundamental concept that underpins our understanding of the natural world. In this chapter, we will look at the essence of frequency and explore its definition; historical context; and pervasive role in nature, technology, the seen realm, and the unseen realm.

DEFINING FREQUENCY

At its core, *frequency* refers to the number of occurrences of a repeating event per unit of time. In physics, it is most commonly used to describe the rate at which a wave oscillates. Frequency is measured in hertz (Hz), with 1 Hz equating to one cycle per second.[1] For instance, a sound wave with a frequency of 440 Hz vibrates 440 times per second, producing the musical note A above middle C.[2]

Frequency is not confined to sound waves alone, however. It also applies to light waves, radio waves, and even the vibrations of atoms. Each type of wave has its own range of frequencies. For example, visible light occupies a frequency range from about 430 trillion to 770 trillion hertz[3], while radio waves range from a few kilohertz to several gigahertz.[4]

HISTORICAL PERSPECTIVES AND KEY THEORIES

The concept of frequency has evolved significantly over time. Ancient cultures were among the first to recognize the rhythmic patterns in nature. The ancient Greeks, for example, observed the periodicity of celestial bodies and developed early models of harmonic motion. The philosopher Pythagoras is credited with discovering the relationship between the length of a vibrating string and the pitch of the sound it produces, laying the groundwork for the study of harmonics.[5]

In the nineteenth century, scientists like Heinrich Hertz and James Clerk Maxwell expanded our understanding of frequency through their work on electromagnetism. Hertz's experiments confirmed the existence of electromagnetic waves and demonstrated that they traveled at the speed of light.[6] Maxwell's equations, formulated in the mid-1800s, describe how electric and magnetic fields oscillate in space, further illuminating the role of frequency in electromagnetic phenomena.[7]

FREQUENCY IN NATURE AND TECHNOLOGY

The significance of frequency extends beyond theoretical physics; it is a fundamental aspect of various natural and technological processes. In nature, frequency manifests in numerous ways. The regular oscillations of a pendulum, the cycle of seasons, and the rhythmic patterns of animal calls are all examples of how frequency plays a role in the natural world.[8]

One of the most striking examples of frequency in nature is the phenomenon of resonance. Resonance occurs when an object vibrates at its natural frequency in response to an external force. A classic example is the way a wine glass can shatter when exposed to a sound frequency that matches its natural resonance. This principle is not limited to glass; it applies to everything from bridges to buildings, underscoring the importance of understanding frequency in engineering and construction.[9]

In technology, frequency is a crucial element in the functioning of various devices. Radios, for instance, rely on frequency to transmit and receive signals. Different stations broadcast at different frequencies, allowing us to tune in to specific channels.[10] Similarly, televisions use a range of frequencies to display images and sound. Modern communication technologies, such as smartphones and wireless networks, operate on a spectrum of frequencies, enabling us to connect and interact in ways that were once unimaginable.[11]

FREQUENCY IN SOUND AND MUSIC

One of the most familiar applications for frequency is sound and music. When we hear a musical note, we are actually perceiving the frequency of the sound wave produced by an instrument or voice. The pitch of the note is directly related to its frequency, with higher frequencies producing higher pitches while lower frequencies result in lower pitches.[12]

Musicians and composers have long understood the power of frequency. The scale of Western music, for example, is based on the harmonic relationships between different frequencies. The interval between notes, such as the octave, is a direct result of these harmonic principles. When a musician plays two notes that are in harmony, they are exploiting the natural resonances of the frequencies involved.[13]

FREQUENCY AND HUMAN PERCEPTION

Our ability to perceive frequency is not limited to the auditory system. The human brain is capable of detecting a wide range of frequencies, from the low rumble of an earthquake to the high-pitched whistle of a dog whistle.[14] This capacity for frequency detection is vital for many aspects of our daily lives, from enjoying music to navigating our environment.

The study of frequency perception extends into other sensory modalities as well. For example, visual frequency, or the frequency of light waves, determines the color we perceive. The visible spectrum ranges from the low-frequency red light to the high-frequency violet light.[15] This interplay between frequency and perception is crucial for understanding how we experience and interact with the world around us.

APPLICATIONS OF FREQUENCY IN MODERN SCIENCE

Modern science continues to explore the applications of frequency in various fields. In medicine, for example, frequencies are used in diagnostic imaging techniques such as ultrasound and magnetic resonance imaging (MRI). These technologies rely on specific frequencies to produce detailed images of the body's internal structures, aiding in diagnosis and treatment planning.[16]

In environmental science, researchers study the frequencies of natural phenomena to better understand and protect ecosystems. For instance, monitoring the frequencies of animal calls can provide insights into population dynamics and habitat changes. Similarly, studying the frequency of seismic waves helps scientists predict and mitigate the impacts of earthquakes.[17]

THE FUTURE OF FREQUENCY RESEARCH AND MEDICAL APPLICATIONS

As our understanding of frequency deepens, new possibilities and applications are continually emerging. Advances in technology and research are likely to unlock even more potential uses for frequency in fields ranging from communication to health. The study of frequency is poised to play a central role in shaping the future of science and technology, revealing new dimensions of our world and expanding our ability to harness its power.

CONCLUSION

Frequency is a concept that permeates every aspect of our lives. From the vibrations of atoms to the waves of communication technology, frequency is a fundamental element of the universe. By exploring the history, science, and applications of frequency, we gain a deeper appreciation for its role in shaping our world and influencing our daily experiences. As we continue to unravel the mysteries of frequency, we stand on the brink of new discoveries that promise to transform our understanding of the universe and our place within it.

PERSONAL REFLECTION

These questions encourage reflection on the intersection of science, faith, and the intricate design of the universe, leading to a deeper appreciation for God's creation and our role within it.

1. What is the significance of frequency in our understanding of the natural world? In what ways does the human ability to perceive different frequencies point to the wonder of God's creation? How does our sensory experience help us connect with the Creator?

2. This chapter mentions resonance, where an object vibrates at its natural frequency in response to an external force. How might this principle of resonance serve as a pathway for our relationship with God? How do we "resonate" with God's voice and purpose in our lives?

3. How does frequency affect our human perception, such as in hearing sounds or seeing colors?

4. How should Christians approach scientific advancements, especially those related to frequencies, in light of biblical teachings about God's ongoing creation and human stewardship of the earth?

5. Consider areas in your own life where you feel out of tune with God's will or purpose. What steps can you take to "align" yourself with His frequency, seeking harmony in your relationship with Him?

CHAPTER 3

THE SCIENCE OF VIBRATION

SCOTT OATSVALL

Vibration is a fundamental concept that lies at the heart of many physical and even spiritual phenomena. From the gentle oscillation of a tuning fork, to the seismic activity of Earth's crust, to the spiritual impact that a person feels, vibrations play a crucial role in shaping our understanding of the world. In this chapter, we will explore the principles of vibration and resonance, discuss how frequency influences physical matter, and examine notable case studies that illustrate these concepts.

PRINCIPLES OF VIBRATION AND RESONANCE

Vibration occurs when an object or system oscillates around a central point. This motion can be periodic, meaning it repeats at regular intervals, or it can be random and irregular. The simplest form of vibration is a harmonic oscillator, which oscillates in a smooth, sinusoidal pattern. A classic example is a mass-spring system, where a mass attached to a spring moves back and forth when displaced from its equilibrium position.[1]

The frequency of vibration, or the number of oscillations per second, is a key characteristic of any vibrating system. It determines the pitch of a sound in musical instruments and the color of light in optics. The amplitude, or the maximum displacement from the equilibrium position, determines the intensity or volume of the vibration.

Resonance is a special type of vibration that occurs when an object is driven at its natural frequency. Every object has one or more natural frequencies at which it tends to vibrate most easily. When an external force matches this natural frequency, the object can experience a dramatic increase in amplitude. This phenomenon can be observed in a variety of contexts, from the musical notes produced by a vibrating string to the collapse of a bridge under specific conditions.[2]

HOW FREQUENCY INFLUENCES PHYSICAL MATTER

Frequency has a profound impact on physical matter. For instance, in the realm of sound, the frequency of sound waves determines the pitch we perceive. High-frequency sound waves produce high-pitched sounds, while low-frequency waves create low-pitched sounds. This principle is fundamental to musical scales and tuning systems.[3]

In materials science, frequency plays a crucial role in understanding how materials respond to vibrations. When exposed to certain frequencies, materials can exhibit different properties, such as changes in elasticity or structural integrity. For example, engineers must account for the resonant frequencies of structures like bridges and buildings to prevent catastrophic failures. The Tacoma Narrows Bridge collapse of 1940 is a historical example where resonance led to the bridge's dramatic oscillations and eventual failure.[4]

In biology, frequency influences various physiological processes. For instance, the frequency of brainwave activity can be associated with different states of consciousness. Alpha waves (8–12 Hz) are linked with relaxed, wakeful states, while beta waves (13–30 Hz) are associated with active, focused attention.[5] A recent study done by the Brain Institute concluded that the heart creates up to one hundred times more electromagnetic frequency than the brain. That means heartfelt emotions like love and happiness can be measured in terms of frequency. Understanding these frequencies can be useful in fields like neurofeedback and cognitive enhancement.

CASE STUDIES IN NATURAL FREQUENCIES

Case studies provide valuable insights into how natural frequencies impact various systems. One notable example is the phenomenon of resonance in musical instruments. The guitar, for instance, relies on the resonance of its body to amplify sound. The body of the guitar is designed to resonate at specific frequencies that enhance the sound produced by the vibrating strings.[6]

Another example is the behavior of the human vocal cords. The vocal cords vibrate at frequencies that produce different pitches. The range of these vibrations is influenced by the tension and length of the vocal cords, as well as the shape of the vocal tract. This interplay of factors allows for the wide variety of vocal sounds that humans are capable of producing.[7] Our voices are also unique, and certain emotional frequencies can be detected by the sound of our voice.

THE TWO MOST POWERFUL VOICES IN THE UNIVERSE

The most powerful voice is no doubt God's voice! But the second most powerful voice is our own voice. When we speak or sing, we produce a series of musical notes and chords that reveal aspects of our personality, many of which are hidden deep within our subconscious. This emotional expression through our voice significantly influences our personal and professional relationships, affecting our overall happiness and success. Often, we are unaware of the vast amount of information we convey with our voice each day. The impact of our spoken words is shaped by the underlying programming of our subconscious mind.

The connection between sound and emotions is well documented. For instance, film soundtracks are crafted to evoke specific emotional responses. Skilled composers manipulate our feelings by triggering our nervous system with carefully designed audio cues. This artful approach can stir our emotions, bring us to tears, inspire hope, or even instill fear.

Ignoring the emotional stress and negativity expressed through our voice can jeopardize our health. Up to 80 percent of health issues may be stress-related[8], and many practitioners and providers that I have spoken to believe that there is a large percent of diseases that have an emotional component. This is why God reminds us to not be fearful. Fifty-two times in the Old Testament God says, "Fear not!" God did not give us a spirit of fear but a spirit of love, power, and a sound mind (2 Timothy 1:7)!

RESONANCE IN ENGINEERING AND DESIGN

In the field of seismology, understanding the frequencies of seismic waves is crucial for earthquake prediction and analysis. Seismic waves travel through Earth's crust and exhibit different frequencies depending on their origin and the geological properties of the area. By analyzing these frequencies, scientists can gain insights into the magnitude and potential impact of earthquakes.[9]

Resonance is a critical consideration in engineering and design. Engineers must account for the natural frequencies of structures to ensure their stability and safety. For example, the design of skyscrapers incorporates damping systems to mitigate the effects of wind-induced resonance.[10] These systems help to dissipate energy and prevent excessive swaying that could compromise the building's integrity.

In automotive engineering, resonance is considered in the design of vehicle components to minimize noise and vibration. For instance, the design of engine mounts and exhaust systems often involves tuning to avoid resonant frequencies that could cause uncomfortable vibrations or noise.[11]

FREQUENCY AND TECHNOLOGICAL INNOVATIONS

Advances in technology continue to leverage the principles of frequency and vibration in innovative ways. In telecommunications, frequency modulation (FM) is a technique used to encode information in a carrier wave by varying its frequency. This technique is widely used in radio broadcasting, where different stations transmit at different frequencies to avoid interference.[12]

In medical technology, techniques such as ultrasound imaging use high-frequency sound waves to produce images of the inside of the body. By analyzing the frequencies of the returning sound waves, medical professionals can obtain detailed images of tissues and organs, aiding in diagnosis and treatment.[13]

There are also great advancements in understanding frequencies in quantum biology. Naturopathic and holistic practitioners and providers are using biological testing and DNA map coding that uses applied integrative biophysics and electromagnetic signals as a noninvasive method for gaining valuable information about the body's most vital functions. These unique types of testing can identify root-cause imbalances that are impacting the body's cellular health. This technology combines the frequencies of voice, bioenergetic feedback, physiology, and quantum physics.

The primary objective is to disclose patterns of stressed or weakened organs, glands, and body systems, and customize healing at a cellular level. These types of testing can also provide valuable data about nutritional profile, nutrition deficiencies, environmental stressors, allergies, hormone imbalances, structural misalignments, food sensitivities, oral health, heavy metals, mold, toxic load, pathogens, and much more.

FUTURE DIRECTIONS IN VIBRATION RESEARCH

As research into vibration and frequency continues to evolve, new applications and technologies are likely to emerge. Advances in materials science and quantum biology may lead to the development of new materials with unique vibrational properties, offering new possibilities in fields ranging from aerospace to healthcare to nanotechnology.[14] Additionally, further exploration of resonance and vibration could enhance our understanding of natural phenomena and lead to new innovations in engineering and design.

CONCLUSION

The science of vibration and resonance is a rich and multifaceted field that intersects with many aspects of our lives. From the musical notes we enjoy to the engineering marvels we rely on, vibration and frequency play a crucial role in shaping our world. By exploring these principles, we gain a deeper appreciation for the forces that influence the physical and technological systems around us and uncover new opportunities for innovation and discovery.

PERSONAL REFLECTION

These questions invite a holistic reflection on the principles discussed in the chapter, merging scientific concepts of vibration and frequency with spiritual insights from Scripture. They should encourage us to explore how both the physical and spiritual realms are interconnected, particularly through the influence of sound and resonance.

1. In Genesis 1:1–3, God speaks the world into existence. Reflect on the power of God's voice as described in this passage. How might this relate to the concept of vibration and resonance, where sound has the power to create and transform?

2. Psalm 29:3–9 describes the voice of the Lord as powerful and transformative. How do the characteristics of God's voice (as described in this psalm) align with the principles of resonance and vibration mentioned in the chapter?

3. This chapter mentions that the heart generates more electromagnetic frequency than the brain. Proverbs 4:23 tells us, "Above all else, guard your heart, for everything you do flows from it." How might the physical and emotional frequencies of the heart influence our spiritual health, relationships, and actions?

4. James 3:5–10 speaks about the power of the tongue to both build up and destroy. How does the idea of frequency in sound and vibration relate to the influence our words have on others? How can we ensure our words bring life rather than harm?

5. This chapter touches on medical applications of sound waves, such as ultrasound imaging. Isaiah 38:21 mentions Hezekiah's healing through a poultice of figs. Do you think God's healing power might work through natural principles such as vibration and frequency? How does this shape your understanding of divine healing?

6. This chapter notes that up to 80 percent of health issues are stress-related, and God commands His people to "Fear not!" repeatedly in the Bible. Reflect on 2 Timothy 1:7: "For God gave us a spirit not of fear but of power and love and self-control" (ESV). How does fear influence our personal vibrations (emotions, health, relationships)? How can we "tune" our frequency to align with God's peace?

CHAPTER 4

FREQUENCY IN THE HUMAN BODY

SCOTT OATSVALL

The human body is a complex system where frequency plays a pivotal role in cellular health, which is responsible for maintaining function at optimal levels. From the rhythms of our heart to the frequencies that influence our brainwaves, understanding how frequency impacts our biology can lead to improved health practices and therapeutic techniques. Frequencies can determine how biological rhythms, health interventions, and techniques like sound therapy intersect with our physiological processes.

BIOLOGICAL RHYTHMS AND INTERNAL FREQUENCIES

Biological rhythms, or circadian rhythms, are natural cycles that regulate various physiological processes in the human body. These rhythms operate on a roughly twenty-four-hour cycle and are influenced by external cues like light and temperature. The circadian rhythm governs sleep-wake cycles, hormone release, and body temperature, affecting our overall health and daily functioning.[1]

Beyond circadian rhythms, our bodies exhibit other internal frequencies. For instance, brainwaves are categorized into different types based on their frequency. Alpha waves (8–12 Hz) are associated with relaxation and calm, while beta waves (13–30 Hz) are linked with alertness and cognitive engagement. Theta waves (4–8 Hz) are observed during light sleep and meditation, and delta waves (0.5–4 Hz) occur during deep sleep.[2] These frequencies play a crucial role in our mental state and cognitive function.

THE IMPACT OF FREQUENCY ON CELLULAR HEALTH AND WELLNESS

The interaction between frequency and cellular health extends to various therapeutic applications. One well-known example is sound therapy, which uses specific frequencies to promote healing and relaxation. Binaural beats, for example, involve playing two slightly different frequencies in each ear, which the brain perceives as a single new frequency. Research suggests that binaural beats can enhance relaxation, improve focus, and reduce anxiety.[3]

Another application is vibroacoustic therapy, which involves applying low-frequency sound waves to the body. This method has been explored for its potential benefits in pain relief, muscle relaxation, and even improved circulation. The vibrations stimulate the body's tissues and can influence physiological responses, potentially aiding in the management of conditions such as fibromyalgia and chronic pain.[4]

TECHNIQUES TO HARNESS BENEFICIAL FREQUENCIES

Several techniques harness the power of frequency to improve cellular health and well-being. Prayer, meditation, and mindfulness practices often incorporate sound and vibration, such as Tibetan singing bowls and tuning forks. These tools use specific frequencies to enhance relaxation and meditative states. Research into these practices suggests that they can reduce stress, improve mood, and enhance overall mental health.[5]

In addition, the field of bioenergetic feedback uses frequency-based technology to help individuals gain greater control over physiological functions. Bioenergetic feedback devices measure physiological parameters such as heart rate variability, brainwave patterns, and even the most sensitive information of a person's vital systems in the body, providing real-time feedback that can help individuals learn to regulate their stress responses and improve their overall cellular health.[6]

CASE STUDIES AND APPLICATIONS

Several case studies highlight the practical applications of frequency-based health interventions. For instance, research on binaural beats has demonstrated their potential to improve sleep quality and cognitive performance. One study found that participants who listened to binaural beats experienced improved sleep and reduced symptoms of insomnia compared to those who didn't use the technique.[7]

Similarly, vibroacoustic therapy has shown promise in various clinical settings. For example, a study involving patients with chronic pain found that those who received vibroacoustic therapy reported significant reductions in pain and improved quality of life. The therapy's low-frequency vibrations were thought to enhance blood flow and reduce muscle tension, contributing to the observed benefits.[8]

There are also technologies such as the Subtle Energy Frequency Imprinter, or S.E.F.I. This technology is designed to capture, amplify, and imprint subtle energy frequencies, and can either broadcast or imprint those frequencies into a person, element, or item. One example of this technology is the A.O. scan by Solex.

FUTURE DIRECTIONS IN FREQUENCY-BASED HEALTH RESEARCH

As research into frequency and health continues, new applications and technologies are likely to emerge. Advances in neurotechnology, such as brain neurofeedback, brainwave modulation, and transcranial magnetic stimulation (TMS), offer exciting possibilities for therapeutic interventions and future health treatments. These technologies may provide new ways to influence brain activity and treat neurological conditions such as TBI and PTSD.[9]

Furthermore, the integration of frequency-based therapies with some traditional practices holds promise for enhancing patient care and evidence-based outcomes. For instance, combining sound therapy with conventional treatments could improve outcomes for conditions such as anxiety, depression, and chronic pain.[10] Continued research and innovation in this field will be crucial for realizing the full potential of frequency-based health interventions.

CONCLUSION

The study of frequency in the human body reveals a rich interplay between physiological rhythms and health interventions. By exploring biological rhythms, frequency-based therapies, and emerging technologies, we will continue to gain valuable insights into how frequency influences our cellular health well-being and opens new avenues for improving health and quality of life.

PERSONAL REFLECTION

1. Reflect on the idea that God created the human body with such intricate design. Psalm 139:13–14 reminds us, "For you created my inmost being; you knit me together in my mother's womb." How does this highlight God's intentionality in our biological makeup?

2. Circadian rhythms, which regulate sleep-wake cycles, are crucial for our health. In Genesis 2:2–3, God rested on the seventh day after creation. How does the principle of rest in Scripture connect with the importance of biological rhythms in maintaining health and balance?

3. Reflecting on the ministry of Jesus in healing the sick (e.g., Matthew 8:16–17), how can we apply the understanding of healing frequencies to our own spiritual and physical well-being?

4. The Bible speaks of peace and calm being central to the Christian life (see Philippians 4:6–7). How can the relaxation benefits of alpha brainwaves and practices like meditation and prayer enhance our spiritual lives and lead to greater peace?

5. This chapter mentions how frequencies can impact cognitive functions, such as focus and alertness. In Colos-

sians 3:2, we are instructed, "Set your minds on things above, not on earthly things." How can this directive be tied to the concept of mental focus, and how might frequency-based therapies support a believer's ability to focus on God's Word?

6. This chapter discusses the potential of new technologies, such as neurofeedback and brainwave modulation, to aid in healing. How do we balance the use of modern technologies in healthcare with our faith and trust in God's power to heal, as seen in James 5:14–15?

7. First Corinthians 6:19–20 teaches that our bodies are temples of the Holy Spirit. How does the idea of maintaining cellular health through the use of frequencies fit into this biblical view of honoring our bodies?

CHAPTER 5
FREQUENCIES AND EMOTIONS

SCOTT OATSVALL

The relationship between frequency and emotions is a fascinating area of study that bridges neuroscience, psychology, therapeutic practices, and even spirituality. In this chapter we'll look at how different frequencies affect our emotional state, explore the mechanisms behind these effects, and discuss practical applications for emotional healing through frequency therapy.

THE ROLE OF FREQUENCY IN EMOTIONAL STATES

Emotions are complex responses to internal and external forces, and frequency plays a significant role in how we experience and process these emotions. For instance, certain sound frequencies can evoke specific emotional responses. Low-frequency sounds, such as the rumble of thunder or the bass in music, are often associated with feelings of fear or awe, while high-frequency sounds, like birdsong or high-pitched music, can induce feelings of joy and calm.[1]

Research into the effects of sound frequencies on emotions has led to the development of various therapeutic practices and treatments in natural healthcare. For example, studies have shown that listening to music with specific frequencies can alter mood and emotional states. Music therapy, which incorporates different musical frequencies and rhythms, is used to help individuals manage stress, anxiety, and depression.[2]

MECHANISMS BEHIND FREQUENCY-DRIVEN EMOTIONAL CHANGES

The mechanisms by which frequency influences emotions are complex and involve both physiological and psychological processes. One key factor is the brain's response to sound frequencies. Different frequencies can activate various brain regions associated with emotion and stress regulation. For example, low-frequency sounds can stimulate the amygdala, a brain region involved in emotional processing, potentially eliciting a stronger emotional response.[3]

Additionally, the autonomic nervous system, which regulates physiological responses such as heart rate and respiration, is sensitive to frequency-based stimuli. Certain frequencies can trigger the parasympathetic nervous system, promoting relaxation and reducing stress. This is one reason why calming music or sounds can be effective in managing anxiety and promoting a sense of well-being.[4]

PRACTICAL APPLICATIONS FOR EMOTIONAL HEALING

Frequency-based therapies have been explored for their potential to facilitate emotional healing. One prominent example is sound therapy, which uses various frequencies to address emotional and psychological issues. Techniques such as binaural beats, brainwave entrainment, and tuning forks are used to help individuals achieve emotional balance and mental clarity.[5]

Another application is the use of specific musical frequencies to enhance emotional well-being. Music therapy often involves selecting music with particular frequencies and rhythms to address emotional needs. For example, slow, calming music with low frequencies may be used to reduce anxiety and promote relaxation, while more upbeat music with higher frequencies can energize and uplift mood.[6]

CASE STUDIES AND SUCCESS STORIES

Several case studies illustrate the effectiveness of frequency-based interventions for emotional healing. For instance, research on binaural beats has shown that listening to these beats can lead to reductions in anxiety and improvements in emotional well-being. One study found that participants who listened to binaural beats experienced lower levels of anxiety and increased relaxation compared to a control group.[7]

Music therapy has also demonstrated significant benefits for emotional health. In a study involving individuals with depression, music therapy sessions that incorporated specific frequencies and rhythms led to reductions in depressive symptoms and improvements in overall mood.[8] These findings highlight the potential of frequency-based interventions to support emotional healing and mental health.

FUTURE DIRECTIONS IN FREQUENCY AND EMOTION RESEARCH

The field of frequency and emotion research, especially in the field of inner voice technologies, is evolving, with ongoing studies exploring new applications and treatments. Advances in neuroimaging and brainwave monitoring are providing deeper insights into how different frequencies affect emotional processing and regulation.[9] These developments may lead to more targeted and effective frequency-based therapies for emotional issues.

Additionally, the integration of frequency-based interventions with other therapeutic modalities holds promise for enhancing emotional healing. Combining sound therapy with mindfulness practices or cognitive-behavioral techniques could offer a comprehensive approach to managing emotional health and improving overall wellness.[10]

THE POWER OF SOUND AND FREQUENCY IN SCRIPTURE

The power of sound and frequency is described in several places in Scripture.

- **Psalm 98:4–6:** "Shout for joy to the LORD, all the earth, burst into jubilant song with music; make music to the Lord with the harp, with the harp and the sound of singing, with trumpets and the blast of the ram's horn—shout for joy before the LORD, the King." This passage demonstrates how sound, music, and frequencies were part of worship in the Bible. The act of music, which is composed of varying frequencies, was used to express joy, gratitude, and praise to God. How do you think different types of music impact your emotional state during worship?

- **Colossians 3:16:** "Let the message of Christ dwell among you richly as you teach and admonish one another with all wisdom through psalms, hymns, and songs from the Spirit, singing to God with gratitude in your hearts." In this verse, Paul encourages believers to use music and song to nurture the message of Christ and enrich the community. What role do you think music and sound play in cultivating peace, joy, and emotional health within the church?

- **1 Samuel 16:23:** "Whenever the spirit from God came on Saul, David would take up his lyre and play. Then relief would come to Saul; he would feel better, and the evil spirit would leave him." Here we see how David's music, played at a particular frequency, had a healing effect on

King Saul's emotional state. What does this tell us about the potential of music and sound in spiritual and emotional healing?

- **2 Chronicles 5:13–14:** "The trumpeters and musicians joined in unison to give praise and thanks to the Lord. . . . Then the temple of the LORD was filled with a cloud, and the priests could not perform their service because of the cloud, for the glory of the LORD filled the temple of God." In this passage, the harmonious sounds and frequencies of musical instruments invited God's presence into the temple. How can music, when played in unity and worship, create a spiritual atmosphere of healing and transformation?

CONCLUSION

The interplay between frequency and emotions reveals a dynamic and impactful relationship. By understanding how different frequencies influence emotional states and exploring practical applications for emotional healing, we gain valuable insights into enhancing mental health and promoting emotional well-being.

PERSONAL REFLECTION

1. Research has shown that different sound frequencies have a direct impact on our emotional states. Low frequencies, such as the rumbling of thunder, often invoke fear or awe, while high frequencies, such as birdsong or uplifting music, can inspire joy or peace. These effects occur because certain frequencies activate specific areas in the brain associated with emotions. How do you experience different emotional responses to sound in your own life? Have you ever noticed how a particular type of music can either calm your heart or make you feel more anxious?

2. Just as David played the lyre to bring healing to Saul, music has been used throughout history to alleviate emotional distress. Music therapy, which uses specific frequencies and rhythms, has shown promise in helping individuals manage anxiety, stress, and depression. Have you ever felt emotionally lifted or healed after listening to specific music? How do you think God might use music as a means of emotional and spiritual healing in your life?

3. In the world of therapy, sound frequencies are used to promote emotional healing. Techniques like binaural beats, which use two different sound frequencies in each ear, help synchronize brainwaves and induce states of relaxation and focus. Similarly, tuning forks and brainwave entrainment are used to restore emotional balance. Have you ever tried any form of sound therapy or music therapy? What was

your experience, and how did it affect your emotional well-being?

4. Sound therapy can be paired with prayer, meditation, and Scripture reading to enhance emotional and spiritual healing. Just as the Israelites used music to invite God's presence, we can use sound in our personal lives to cultivate peace and draw near to God. How can you incorporate music or sound frequencies into your spiritual practices to enhance your emotional and spiritual well-being?

5. As research into the effects of frequency on the brain and emotions advances, there is potential to integrate sound therapies with spiritual practices for holistic healing. By combining scientific insights with biblical principles, we can explore new ways to use sound to strengthen our emotional and spiritual health. How do you see the relationship between science and faith in the area of emotional healing? How can both work together to enhance our well-being, or are they separate realms?

CHAPTER 6
FREQUENCIES IN COMMUNICATION

SCOTT OATSVALL

Frequency is a fundamental aspect of communication, influencing how we transmit and receive information across various platforms. From radio waves to digital signals, the role of frequency in communication technologies is essential for understanding our connections and interactions in the modern world. This chapter delves into the principles of frequency modulation, the impact of frequency on interpersonal communication, and strategies for enhancing communication through frequency awareness.

FREQUENCY MODULATION IN TECHNOLOGY

Frequency modulation (FM) is a technique that encodes information into a carrier wave by varying its frequency. This method is prevalent in radio broadcasting, where different stations transmit at distinct frequencies to minimize interference. FM allows for high-quality sound transmission and is less susceptible to signal degradation compared to amplitude modulation (AM).[1] The inherent stability of FM makes it a preferred choice for numerous audio applications, including music and talk shows, ensuring that listeners receive clear and consistent broadcasts.

Beyond radio, frequency modulation is integral to various technologies. In television broadcasting, FM helps transmit audio and video signals with remarkable fidelity. The clarity provided by frequency modulation allows viewers to experience rich sound alongside high-resolution images, significantly enhancing their viewing experience.[2] Moreover, FM plays a crucial role in wireless communication systems such as cellular networks and Wi-Fi. These technologies rely on frequency modulation to transmit data efficiently, facilitating seamless connectivity in an increasingly digital world.

For example, in a typical mobile phone conversation, frequency modulation ensures that the voice of the caller is transmitted clearly, allowing for effective communication regardless of distance. The same principles apply to Wi-Fi networks, where FM enables data to be sent rapidly and accurately between devices. This capability has transformed the way we interact, work, and socialize, highlighting the importance of frequency in our daily lives.

THE IMPACT OF FREQUENCY ON INTERPERSONAL COMMUNICATION

Frequency also significantly affects interpersonal communication, particularly in how we perceive and interpret spoken and nonverbal cues. The pitch of a person's voice, determined by the frequency of their vocal cords' vibrations, conveys essential information about their emotional state and intent.[3] Higher frequencies are often associated with excitement, joy, or love, while lower frequencies may indicate stress, anger, or sadness. This relationship between voice frequency and emotional expression underscores the nuanced ways we communicate and the importance of vocal modulation in conveying meaning.

In addition to vocal frequency, nonverbal communication cues such as body language and facial expressions play a vital role in interpersonal interactions. The frequency of these nonverbal signals can influence how we interpret messages. For instance, a person who leans in and speaks at a higher pitch may be expressing enthusiasm or urgency, while someone who adopts a closed posture and uses a lower tone may be signaling discomfort or disinterest. By being attuned to these frequency-based cues, individuals can enhance their understanding of others' emotions and intentions, fostering more effective communication.

STRATEGIES FOR ENHANCING COMMUNICATION THROUGH FREQUENCY AWARENESS

Recognizing the significance of frequency in both technological and interpersonal communication opens avenues for improving our interactions. Here are several strategies to enhance communication through frequency awareness:

- **Vocal Awareness:** Practicing vocal modulation can improve clarity and emotional expression in spoken communication. By varying pitch, tone, and volume, individuals can convey their feelings more accurately, making their messages more resonant with listeners. For example, during a presentation, varying one's vocal frequency can capture the audience's attention and emphasize key points.

- **Active Listening:** Engaging in active listening involves paying attention to not only the words spoken but also the frequency and tone used by the speaker. This attentiveness helps to decode underlying emotions and intentions, allowing for a more empathetic response. By reflecting back on what was heard, including frequency cues, individuals can demonstrate understanding and build stronger connections.

- **Nonverbal Signals:** Being mindful of nonverbal communication frequency can enhance interpersonal interactions. Body language, such as posture and gestures, conveys frequency-related messages that complement spoken words. For instance, maintaining an open posture while speaking can signal receptiveness and engagement, while mirroring the other person's body language can create a sense of rapport.

- **Feedback Mechanisms:** Creating environments where feedback is encouraged can enhance communication frequency awareness. This can include regular check-ins in professional settings or open discussions about communication styles in personal relationships. By understanding how frequency influences perceptions, individuals can adjust their communication methods to better suit their audience.

- **Technological Utilization:** Leveraging technology to facilitate better communication can also be beneficial. For instance, utilizing software that analyzes vocal frequency can provide insights into how effectively one is communicating. This can help individuals recognize patterns in their speaking styles and make necessary adjustments to improve clarity and emotional resonance.

CONCLUSION

Frequency is a pivotal component of both technological and interpersonal communication. By understanding the principles of frequency modulation, we can appreciate the technological advancements that enable our connectivity. Additionally, recognizing the impact of frequency on our vocal and nonverbal cues allows us to refine our communication skills, fostering deeper connections and more meaningful interactions. As we continue to navigate a rapidly evolving communication landscape, embracing frequency awareness will undoubtedly enhance our ability to connect and collaborate effectively.

PERSONAL REFLECTION

1. Just as FM provides clearer communication and avoids interference, we as Christians are called to speak clearly, without distortion or confusion. Our words should not be muddled with negativity, gossip, or bitterness, but rather should reflect God's truth and love. Colossians 4:6 tells us, "Let your conversation be always full of grace, seasoned with salt, so that you may know how to answer everyone." How does the concept of frequency modulation in technology—changing frequencies to prevent interference and ensure clear communication—relate to the way we communicate with others in our relationships? Is there a "frequency" at which our communication is clearer and more effective?

2. The Bible speaks about the importance of the heart in communication. Jesus said that "the mouth speaks what the heart is full of" (Matthew 12:34). If our hearts are filled with peace, patience, and kindness, our words will reflect that frequency. Conversely, if our hearts are filled with anger or bitterness, our words will carry that negative energy as well. "The fruit of the Spirit is love, joy, peace, forbearance, kindness, goodness, faithfulness, gentleness, and self-control" (Galatians 5:22–23). How do the "frequencies" of your tone, volume, and body language affect the messages you are trying to convey? Do you communicate with love, gentleness, and respect, or do you sometimes communicate frustration or impatience?

3. The Bible emphasizes that we should speak words that build up rather than tear down: "Let everything you say be good and helpful, so that your words will be an encouragement to those who hear them" (Ephesians 4:29 NLT). Jesus also taught us to speak with love and to turn our hearts toward reconciliation: "Blessed are the peacemakers, for they will be called children of God" (Matthew 5:9). Awareness of the "frequency" at which we communicate can improve the effectiveness of our interactions. Just as tuning a radio to the correct frequency results in clear reception, tuning our communication to the right "frequency" can help us relate better to others. Are you aware of the emotional or spiritual "frequency" at which you communicate? How might you need to adjust your tone, words, or body language to communicate more effectively with others, especially in stressful situations?

4. Jesus is the ultimate communicator. He spoke with authority and love, and His words brought healing, comfort, and truth. He called His followers to a higher standard of communication, emphasizing that our words should reflect the character of God: "For by your words you will be acquitted, and by your words you will be condemned" (Matthew 12:37). Just as radio signals can carry different types of information depending on the frequency, our words can carry different kinds of messages—either life-giving or destructive. We are called to use our words to reflect the love of Christ and to share the good news. How can we use our words to build up the body of Christ? How

do we ensure that the "signal" we are transmitting through our words and actions aligns with God's truth?

CHAPTER 7
FREQUENCIES AND THE HOLY SCRIPTURES

SCOTT OATSVALL

The concept of frequency, while often discussed in medical, scientific. and technological contexts, also finds intriguing parallels, and I would submit is rooted in biblical and spiritual references. Holy scriptures from various cultures sometimes refer to principles that resonate with the notion of frequency, though not always in explicit terms. This chapter explores how frequency-related references appear in holy scriptures, examining their symbolic and metaphorical meanings, and discussing their implications for people of faith.

FREQUENCY IN ANCIENT TEXTS AND SYMBOLISM

The Bible as well as other ancient texts speak of frequency quite often. For example, the concept of vibration or resonance can be seen in the beginning of the Bible in Genesis 1:2: "Now the earth was formless and empty, darkness was over the surface of the deep, and the Spirit of God was hovering over the waters."

This word *hovering* is the Hebrew word *rāḥap̄*, which means to move or to shake—or in other words, to vibrate! This is very important, because quantum physics has taught us in modern-day science terms that vibration holds all matter together. The significance is that God was putting His frequencies into the earth as He was forming the earth.

There are many biblical references to sound and vibration that can be interpreted through the lens of frequency. For instance, the book of Psalms frequently speaks of praising God with musical instruments and voices. Psalm 150 calls for the use of various instruments, including trumpets, harps, and cymbals, to glorify God. The use of music and sound in worship is seen as a means of connecting with the divine and expressing reverence.[1]

Another notable example is the story of the fall of Jericho in the Book of Joshua. The Israelites, instructed by God, marched around the city blowing trumpets. On the seventh day, after circling the city seven times, the priests blew their trumpets, and the walls of Jericho fell down. This event suggests the use of sound and vibration as a divine mechanism for achieving miraculous outcomes.[2] While the Scriptures don't use the English word *frequency*, this narrative can be interpreted as an example of how sound and vibration (frequencies) are employed to effect physical and spiritual change.

ESOTERIC AND MYSTICAL INTERPRETATIONS

Esoteric traditions often explore the concept of frequency in more abstract and mystical terms. In Kabbalah, a form of Jewish mysticism, the Tree of Life and its associated *sefirot* (divine emanations) are sometimes interpreted in terms of vibrational frequencies. Each *sefirah* is thought to represent a different aspect of divine energy, and the interactions between them can be seen as a form of cosmic resonance.[3]

In Christian theology, the idea of divine "frequency" can be linked to the concept of divine harmony and order. The idea that God created the universe with an inherent order and beauty resonates with the notion of a cosmic frequency that governs all creation. This perspective aligns with the view that spiritual practices, such as prayer and worship, help individuals align themselves to God almighty.

The Bible, as you can see, contains numerous references to sound, music, and vibrations, which can be interpreted through the concept of frequency. While the term *frequency* is not explicitly used in biblical texts, the underlying principles of sound and vibration play a significant role in the spiritual and ritual life described in the Scriptures.

THE SYMBOLISM OF SOUND AND INSTRUMENTS IN BIBLICAL NARRATIVES

Sound and music are integral to various rituals and practices described in the Bible. The use of musical instruments and vocal expressions in worship is prominent, reflecting an understanding of sound's role in connecting with God. The diverse range of instruments recorded in the Bible underscores the importance of sound in worship, suggesting a form of frequency that enhances spiritual engagement.

Biblical narratives often use sound and vibration

physically and symbolically to convey divine messages or interventions. The sound of thunder and the voice of God are prominent in key moments of the Scriptures. For example, in the book of Exodus, God speaks to Moses from Mount Sinai amid thunder and lightning, creating an awe-inspiring environment that underscores the divine presence.[4] The thunder and lightning can be interpreted as symbolic frequencies representing God's power and majesty, emphasizing the profound impact of divine communication.

Another significant instance is the voice of God described in the book of 1 Kings. Elijah experiences God's presence not through dramatic manifestations such as wind, earthquakes, or fire, but through a "still small voice."[5] This subtle sound contrasts with more dramatic manifestations and suggests a different form of divine communication that operates at a more profound frequency, emphasizing the importance of quiet and inner alignment as people experience God at a deeper level.

Music and chanting play a critical role in the biblical tradition, contributing to spiritual and communal experiences. The use of music in worship, as depicted in the Psalms, involves a rhythmic and harmonic quality that can be associated with frequency. The act of singing praises to God with a harmonious voice is seen as a way to align with divine frequencies, enhancing spiritual connection and worship.[6]

The story of King David provides another example of how music is used to influence spiritual and emotional states. David's harp playing is described as soothing to King Saul, who was tormented by an evil spirit.[7] This narrative suggests that music, through its frequency and vibration, has the power to impact the emotional and spiritual well-being of individuals, offering comfort and healing as well as power over the spiritual forces of darkness.

BIBLICAL TEXTS AND THE CONCEPT OF DIVINE HARMONY

The Bible's descriptions of divine order and harmony can be viewed through the lens of frequency. The concept of a divinely ordered universe is reflected in the way creation is described. In Genesis, the creation of the world involves God's spoken word bringing order and form to the universe. The phrase "And God said, 'Let there be light'" reflects the power of divine speech, which can be interpreted as a form of frequency that initiates creation and transformation.[8] This is why your voice is also very powerful. **WHAT YOU SPEAK MATTERS!**

Additionally, the notion of divine harmony is present in the way the biblical cosmos is described as operating in accordance with God's will. The natural order and the regularity of natural phenomena, such as the cycles of day and night, can be seen as manifestations of a divine frequency that governs creation.[9] This perspective aligns with the idea that the universe operates according to certain principles or "frequencies" established by God.

CONTEMPORARY INTERPRETATIONS AND APPLICATIONS

In contemporary spiritual practices, there is a growing interest in exploring the relationship between biblical principles and the concept of frequency. Modern interpretations often seek to bridge ancient texts with contemporary understandings of sound and vibration. For instance, some practitioners draw on biblical references to sound and music to support the use of sound therapy and frequency-based healing techniques. They interpret biblical passages as evidence of the spiritual significance of sound and vibration in divine communication and healing.

However, it is important to approach these interpretations with discernment. While the metaphorical use of frequency in biblical texts provides valuable insights, it is essential to distinguish between scriptural symbolism and modern scientific concepts. The integration of biblical principles with contemporary practices should be done thoughtfully, respecting both the traditional context of the Scriptures and the empirical nature of modern science.[10]

CONCLUSION

The exploration of sound and vibration in biblical texts reveals a profound connection between spiritual practices and the concept of frequency. From the use of musical instruments in worship to the symbolic representation of divine communication, the Bible reflects an understanding of sound's role in connecting with the divine and manifesting spiritual realities. By examining these biblical references through the lens of frequency, we gain deeper insights into the spiritual significance of sound and its impact on worship, healing, and divine interaction.

PERSONAL REFLECTION

1. In Genesis 1:2, the Spirit of God is described as "hovering" over the waters. How does understanding this as vibration or resonance change your perspective on God's creative act in the universe?

2. How can we apply the idea that God's spoken word (Genesis 1:3) operates as a divine frequency in our lives today, and what does this teach us about the power of speech and prayer?

3. How might our understanding of sound and music enhance our spiritual life and deepen our worship?

4. In 1 Kings 19:11–12, God speaks in a "still small voice." How can we apply this idea of tuning into God's frequency to our own lives?

5. The Bible speaks of God creating the world with order and harmony. How might the concept of divine frequencies help us understand the regularity and beauty of God's creation?

6. How does recognizing God's creation as operating in harmony and frequency shape our understanding of nature, science, and our responsibility as stewards of the earth?

7. With the growing interest in sound therapy and frequency healing, how can we approach these modern practices in light of biblical principles?

CHAPTER 8

THE NEW AGE MOVEMENT AND THE HIJACKING OF FREQUENCIES

SCOTT OATSVALL

In exploring the concept of frequencies, it is essential to recognize that God, as the Creator, is the ultimate originator of all things, including the frequencies that permeate our world and our very existence. While the New Age movement has hijacked various interpretations of these concepts, it is crucial for Christians to understand that new information and technologies regarding frequencies should not discourage or distract them from their faith. Instead, these discoveries can be viewed through the lens of the biblical worldview and do not take away from the Gospels, affirming that they do not alter the core truths of Christianity. Embracing this knowledge can deepen our understanding of God's creation and strengthen our faith, encouraging a pursuit of truth that aligns with Scripture.

God is the ultimate Creator of all things, including the forces and frequencies of nature. Genesis 1:1–3 says, "In the beginning, God created the heavens and the earth. Now the earth was formless and empty, darkness was over the surface of the deep, and the Spirit of God was hovering over the waters. And God said, 'Let there be light,' and there was light."

Christ, as the Creator, is the source and sustainer of all frequencies and vibrations. We can see this in Colossians 1:16–17: "For by Him all things were created: things in heaven and on earth, visible and invisible, whether thrones or powers or rulers or authorities; all things have been created through Him and for Him. He is before all things, and in Him all things hold together." The importance of discernment when faced with spiritual practices or ideas that may not align with biblical truth: "Dear friends, do not believe every spirit, but test the spirits to see whether they are from God, because many false prophets have gone out into the world" (1 John 4:1).

A CLOSER LOOK AT THE NEW AGE MOVEMENT

The New Age movement, which emerged in the late twentieth century, has had a profound influence on various aspects of modern spirituality and alternative health practices. One of its central tenets is the use of frequency and vibration as tools for personal transformation and healing. While these concepts have roots in proven scientific principles, the New Age movement has often adopted and reinterpreted them in ways that diverge significantly from their original contexts. This chapter explores how the New Age movement has appropriated the concept of frequency, examines its impact on contemporary practices, and discusses the implications of this phenomenon.

THE NEW AGE MOVEMENT AND FREQUENCY

The New Age movement encompasses a diverse range of spiritual and holistic practices that emerged in the 1960s and 1970s, drawing from Eastern philosophies, occult traditions, and alternative medicine.[1]

Central to many holistic healthcare strategies, the New Age movement also understands the belief that everything in the universe vibrates at specific frequencies, and that altering these frequencies can lead to spiritual and physical healing. This notion often incorporates concepts such as energy fields, chakras, and crystal healing.[2] I would submit to you that this alteration is not always the way God intended for these frequencies to be used to glorify Him. This is the reason it is critical to understand where these technologies and practices originate.

One prominent example is the use of "healing frequencies," which are claimed to resonate with specific parts of the body or mind to promote wellness. Practices such as sound baths, tuning fork therapy, and binaural beats are often employed under the belief that these frequencies can realign the body's energy and enhance spiritual well-being.[3] While these practices draw on the legitimate scientific and medical concepts of frequency, they frequently lack biblical support and can sometimes misrepresent the God of the Bible. This is why you need to be very certain about where the frequencies are coming from and who is administering the frequencies.

Here are some reasons why, without discernment, the use of frequencies can be dangerous to the Christian.

- **Theological Confusion:** New Age beliefs often incorporate elements of spirituality that contradict core Christian doctrines. This can lead to confusion about the nature of God, salvation, and the authority of Scripture.

- **Syncretism:** New Age practices often blend various religious beliefs, which can dilute the distinctiveness of Christianity and lead believers to adopt practices that are incompatible with the Christian faith.

- **Focus on Self:** Many New Age philosophies emphasize self-realization and personal power over reliance on God. This can shift the focus away from Christ and undermine the foundational Christian principle of humility and dependence on divine grace.

- **Occult Practices:** Some New Age healing techniques may involve practices that Christians consider occult, such as channeling or energy work. Engaging with these practices can expose believers to spiritual dangers that conflict with their faith.

- **Misplaced Trust:** Placing faith in alternative healing methods can lead individuals to rely on human solutions rather than divine intervention, potentially causing them to neglect prayer and the sacraments.

In summary, while frequency can be used in the pursuit of healing and well-being, Christians must remain vigilant and discerning, ensuring that their beliefs and practices align with biblical teachings.

THE SCIENTIFIC CONCEPTS OF HEALING

Holistic and naturopathic healthcare has used frequency and bioenergetic feedback as part of healing strategies for decades. For example, the concept of "healing frequencies," such as 432 Hz or 528 Hz, is frequently thought as having special healing properties. Proponents claim that music tuned to these frequencies can harmonize with the body's energy fields and promote healing.[4] While the frequencies themselves are scientifically valid and have been shown to have great value to the healing process of the human body, I believe until the medical and scientific community has rigorous scientific evidence, unfortunately most of the research will be anecdotal.

Similarly, the idea that crystals possess specific frequencies that can influence physical and emotional health is a common assertion in New Age practices. Crystal healing is based on the belief that different crystals emit particular frequencies that can affect the body's energy field. However, while crystals do have physical properties and can vibrate at certain frequencies, there is little scientific evidence to support the idea that these vibrations have a direct impact on health.[5]

IMPACT ON MODERN PRACTICES AND BELIEFS

The New Age reinterpretation of frequency has led to a proliferation of products and services marketed as frequency-based solutions for a wide range of issues. This can undermine the effectiveness of holistic practitioners and naturopathic physicians because of the potential risk for false claims. The popularity of these claims is driven by the appeal of using frequency as a tool for personal transformation and wellness, but it also raises concerns about the potential for misleading claims and misuse of the original design from God.[6]

The commercialization of frequency-based practices also reflects a broader trend within the New Age movement of appropriating scientific terminology to lend credibility to spiritual or alternative health claims. By using scientific-sounding language, marketers can appeal to a sense of rationality and legitimacy, even when the underlying principles lack empirical support.[7] This highlights the need for critical evaluation of such claims and the importance of distinguishing between scientifically grounded practices and those based on speculative or unverified theories. Again, until the mainstream medical and scientific community give validation and credibility, much-needed research will go mostly to the people willing to challenge the narrative of allopathic healthcare.

CASE STUDIES AND CRITICISMS

One example is the belief that the frequency of 528 Hz is the "love frequency," purported to have transformative effects on health and emotional well-being. Many people who have worked with frequency-based therapy suggest that this frequency can repair DNA and enhance overall health. However, the mainstream scientific community will only recognize that 528 Hz is a valid frequency, and says there is no evidence to support the claims made about its healing properties.[8] This disconnect between holistic care and the scientific community underscores the need for more rigorous evaluation of such practices. But to accomplish this, both sides must be willing to open up all data and research and be willing to investigate with open minds and exchange of unbiased ideas.

THE ROLE OF HEALTHY SKEPTICISM AND CRITICAL THINKING

Encouraging critical thinking and scientific literacy is essential for navigating the complex landscape of frequency-related practices. People seeking all-natural and alternative healthcare should be aware of the potential for pseudoscience and misleading claims, and practitioners should be encouraged to base their approaches on sound scientific principles and empirical evidence. By fostering a more informed and discerning approach, as believers we can better appreciate the genuine and positive potential of frequency while avoiding the pitfalls of unfounded claims and going down the road of New Age practice.

CONCLUSION

The New Age movement's appropriation of frequency concepts reflects a broader trend of integrating scientific terminology into spiritual and alternative health practices. While the principles of frequency are grounded in legitimate science, their application within the New Age context often involves speculative claims and a departure from a biblical understanding of the frequencies created by God. By critically examining these practices and promoting scientific literacy and the biblical understanding of frequency, we can better bridge the gap of science and faith and ensure that our understanding of frequency remains grounded in God's Word.

PERSONAL REFLECTION

1. What are some key differences between how the New Age movement views frequencies and how we as Christians should view them?

2. In 1 John 4:1, we are told to test the spirits. What does this mean in the context of evaluating practices like sound baths, crystal healing, and other frequency-based therapies?

3. How can we as believers practice discernment when we encounter new ideas about healing or spirituality that are not grounded in Scripture?

4. Think about how you've personally engaged with New Age practices or concepts in the past (e.g., meditation, crystals, sound therapy). How might a biblical understanding of frequencies shift your approach to these practices?

5. Reflect on the ways in which God has used traditional and non-traditional methods of healing in your life or in the lives of others. How do you differentiate between methods that are aligned with Scripture and those that are not?

CHAPTER 9
USING FREQUENCIES IN ALTERNATIVE MEDICINE

SCOTT OATSVALL

The concept of frequency has become increasingly prominent in alternative medicine, especially holistic healthcare where it is employed in various therapeutic modalities aimed at promoting health and well-being. This chapter explores how frequencies are utilized in alternative medicine, examining the principles behind these practices, their applications, and the evidence supporting their effectiveness.

THE SCIENCE OF FREQUENCIES IN ALTERNATIVE MEDICINE

Frequencies, defined as the number of vibrations or cycles per second, are foundational in various forms of alternative medicine. Proponents of frequency therapies argue that every substance, including the human body, operates at a specific frequency and that disturbances in these frequencies can lead to health issues.[1] The therapeutic goal, therefore, is to restore balance by adjusting these frequencies.

One common application is frequency-specific

microcurrent therapy (FSM). This technique involves applying low-level electrical currents to the body at specific frequencies. The idea is that different frequencies target different tissues or conditions, such as pain or inflammation.[2] FSM practitioners claim that this therapy can enhance cellular repair, reduce pain, and improve function by restoring the body's natural frequency balance.

Another well-known approach is bioresonance therapy, which is based on the principle that every cell emits electromagnetic frequencies. Bioresonance therapy uses devices to measure these frequencies and then aims to correct imbalances by sending back modified frequencies.[3] Proponents believe that this can help diagnose and treat various conditions, including allergies, chronic pain, and even cancer.

SOUND THERAPY AND HEALING FREQUENCIES

Sound therapy, also known as sound healing or vibrational medicine, uses sound frequencies to promote healing and well-being. This approach is based on the idea that sound vibrations can affect the body's energy fields and influence its healing processes.[4] Various forms of sound therapy include

- **Tuning Fork Therapy:** This method uses metal tuning forks that are struck to produce specific frequencies. The forks are placed on or near the body, with the belief that the vibrations help balance energy and alleviate pain.[5] Proponents suggest that different frequencies target different energy pathways, similar to acupuncture but without needles.

- **Binaural Beats:** This technique involves playing two slightly different frequencies in each ear, which the brain perceives as a single, new frequency. The goal is to synchronize brainwave activity to enhance relaxation, focus, or sleep.[6] Some studies suggest that binaural beats may aid in reducing anxiety and stress and improving cognitive performance.

- **Solfeggio Frequencies:** These are specific frequencies associated with ancient musical scales believed to have healing properties. For instance, the 528 Hz frequency is often referred to as the "love frequency" and is claimed to promote healing and DNA repair. 432 Hz is said to provide calmness and balance. 639 Hz is often associated with strengthening connections and relationships, while 852 Hz is said to be linked to returning to spiritual connection to God.[7]

- **Bioenergetic feedback:** Bioenergetic feedback and full body frequency scans look at over 800 panels in the human body that can be done in person or virtually. This is called biological profiling and DNA map coding. Neither distance nor accuracy is compromised, and a full report and specific strategies and recommendations for specific needs is given afterward. The reason this scan is so valuable is that a health crisis is usually preceded by decades of deterioration before a symptom may manifest. Several known root problems are focused on, with the goal being to address any crisis at its root and not just put a Band-Aid on the symptoms. Also included in the scans are inner voice print technology. Skilled bioenergetic holistic practitioners use the energetic information transmitted by the human voice to test for things like trauma, stress, anxiety, anger, truthfulness, and the physical condition to name a few. Visit www.LT360.com for more information on full body scanning and frequency testing.

ELECTROMAGNETIC FIELD THERAPIES

Electromagnetic field (EMF) therapies are based on the principle that the body's health can be influenced by external electromagnetic fields. These therapies use various devices to expose the body to specific frequencies and electromagnetic fields, aiming to improve health and alleviate symptoms.

- **Pulsed Electromagnetic Field Therapy (PEMF):** This involves exposing the body to low-frequency electromagnetic pulses. The therapy is purported to stimulate cellular repair, reduce inflammation, and improve circulation.[8]

Some studies suggest that PEMF may be beneficial for conditions such as osteoarthritis and bone healing, but further research is needed to confirm its efficacy.

- **Transcranial Magnetic Stimulation (TMS):** This uses magnetic fields to stimulate nerve cells in the brain. It is primarily used to treat depression and other neurological conditions by modulating brain activity. While TMS is a well-established treatment in clinical settings, its application in alternative medicine often extends to broader claims about its ability to treat various mental health issues.[9]

The effectiveness of frequency-based therapies in alternative medicine is a subject of ongoing development and study. While some practices, such as TMS and PEMF, have garnered scientific support for specific applications, we need more holistic practitioners and providers to come forth with their clinical outcomes to give further credibility.

Additionally, the lack of standardization and regulation in alternative medicine can lead to variability in treatment practices and outcomes. Consumers should be cautious and seek therapies that are administered and conducted by qualified practitioners.[10] It is also important to know who is administering the scans and if they are Spirit-filled practitioners who are Holy Spirit and Kingdom people!

CONCLUSION

Frequencies play a multifaceted role in alternative medicine, from sound and electromagnetic therapies to bioenergetic feedback and inner voice technology. As the field of alternative and holistic medicine continues to evolve, it is crucial to balance innovative approaches with evidence-based outcomes. By fostering a deeper understanding of how frequencies can impact health and well-being, we can better integrate these practices into holistic health paradigms and ensure that they are used effectively and safely.

PERSONAL REFLECTION

1. If every part of the body has a specific "frequency," how does this understanding deepen your view of how the body is designed for health and healing?

2. How does Exodus 15:26, where God promises healing for His people, align with the idea of restoring the body's natural frequencies?

3. Sound therapy uses vibrations to influence health. In Proverbs 4:20–22, God's Word is described as bringing life and health to the body. How might sound frequencies in therapy reflect this biblical truth?

4. Electromagnetic field therapies, such as PEMF and TMS, are based on the principle that external energy fields can affect the body. In Luke 8:43–48, a woman is healed by the power flowing from Jesus. How can this be understood in terms of energy or frequency?

5. How can we integrate spiritual principles with medical practices, especially in areas like sound healing, frequency therapy, and electromagnetic therapies while maintaining faith in God's healing power as we are gaining understanding in alternative medicine?

CHAPTER 10
THE FREQUENCY OF THOUGHTS

SCOTT OATSVALL

Our thoughts create energy. Each thought we entertain emits a frequency, shaping our reality in profound ways. This energy can be either positive or negative, influencing our emotions, behaviors, and overall well-being.

Scientific studies, particularly in the realm of quantum mechanics, reveal that negative energy is less potent than what exists in empty space. This suggests that when we engage in negative thinking, we produce a lesser energy state, potentially dragging us down.[1] In contrast, positive thoughts resonate with higher frequencies, empowering us and fostering resilience. Neuroscience supports this view, indicating that our thoughts can physically alter the structure of our brains. Minute by minute, each thought contributes to a complex neural landscape known as neuroplasticity.[2] Established patterns of thinking can be deeply entrenched, making it challenging to break free from negative cycles. This chapter explores how our very thoughts generate energy that affects our very existence mentally, physically, emotionally, and spiritually. We will also look at the medical and scientific ways that the frequency of thought as well as genetics affect our very well-being and overall health.

BREAKING FREE FROM ESTABLISHED PATTERNS

To escape these well-established pathways of thought, we must actively create new mental frameworks. Here are several strategies to facilitate this transformation:

- **Learn Something New:** Engaging in new activities or acquiring new knowledge can stimulate different neural connections, effectively reshaping our thought patterns.[3] When we challenge ourselves intellectually, we encourage the brain to forge new pathways.

- **Practice H.A.L.T.:** This acronym stands for Hungry, Angry, Lonely, Tired, and serves as a reminder to check in with ourselves. When we are in these states, our capacity to think positively diminishes, making it harder to break free from negativity. Recognizing these feelings allows us to address them directly and shift our mindset.[4]

- **Understand Motivation:** Change often occurs in response to a perceived threat or the promise of reward. When we associate pleasure with new thoughts and pain with negative ones, we create a compelling reason to shift our mindset.[5]

THE CREB GENE

The role of genes in this process is also critical. The CREB gene, or CREB-regulated transcription coactivator 1 (CRTC1), plays a significant role in regulating gene expression in response to various stimuli, including stress and metabolic factors.[6] It acts as a coactivator for the cAMP response element-binding protein (CREB), which is crucial for learning, memory, and energy balance. In a sense, we can think of it like a light switch that can be turned on or off, influencing our ability to adapt and grow based on our thoughts and experiences.

Key functions of the CREB gene include

- **Gene Regulation:** The CREB gene enhances the transcription of target genes involved in metabolism and neurobiology.[7] This means it helps control how our bodies respond to various stimuli, linking our thoughts to our physical state.

- **Metabolic Processes:** The CREB gene influences energy homeostasis and is associated with conditions such as obesity and diabetes.[8] Understanding this connection reveals how our mental state can impact our physical health.

- **Neuronal Function:** The CREB gene is involved in critical processes for learning and memory, as it modulates synaptic plasticity.[9]

This underscores the profound link between our thoughts and our cognitive abilities.

Mutations or dysregulation of the CREB gene have been linked to metabolic disorders and various neurodegenerative diseases.[10] Some researchers suggest that up to 90 percent of known chronic illnesses may be connected to this gene. This highlights the importance of understanding the frequency and nature of our thoughts, as they can directly affect our health.

THE SPIRITUAL DIMENSION OF THOUGHTS

The Bible also offers profound insights into the significance of our thoughts. For example, Proverbs 23:7 reminds us, "For as he thinks in his heart, so is he" (NKJV). This verse emphasizes that our inner thoughts shape our identity and character. Similarly, Philippians 4:8 encourages us to focus on what is true, noble, right, and pure. This guidance serves as a powerful reminder to cultivate a mindset that aligns with positive values.

Romans 12:2 reinforces this idea, urging us not to conform to negative patterns but to renew our minds continually. This renewal is essential for transformation. Furthermore, 2 Corinthians 10:5 encourages us to take every thought captive, ensuring that our mental landscape is in harmony with our spiritual beliefs.

THE IMPACT OF WORDS

Our thoughts also manifest through our words. Matthew 12:34 states, "For the mouth speaks what the heart is full of." This connection between thoughts and speech underscores the importance of guarding our minds. Psalm 19:14 offers a beautiful prayer, asking for our words and meditations to be pleasing to God, reinforcing the idea that our thoughts should align with God's principles. And Isaiah 26:3 promises peace to those whose minds are steadfast, a reflection of trust in God. This verse highlights that when we focus our thoughts on faith and positivity, we can attain a state of tranquility, regardless of external circumstances.

CONCLUSION

Before thoughts become words or actions, they first emerge in the mind. This foundational understanding reveals the profound power of thought frequency. By consciously choosing our thoughts, we can reshape our reality; cultivate positivity; and ultimately influence our mental, physical, and spiritual health.

As we navigate the complexities of life, let us be mindful of the energy and frequency we create through our thoughts. By embracing new ways of thinking and aligning our minds with the Holy Spirit, we can unlock our full potential and experience a richer, more fulfilling life.

PERSONAL REFLECTION

1. What do you think Proverbs 23:7 means when it says, "For as he thinks in his heart, so is he" (AMP)? How does this relate to the concept of thoughts emitting a frequency or energy?

2. Can you think of a time when your thoughts—whether positive or negative—shaped your experience or behavior in a significant way? How did your mindset influence the situation?

3. Romans 12:2 speaks about the importance of renewing the mind. What practical steps can you take to transform your thinking patterns? How can this renewal lead to spiritual growth?

4. Philippians 4:8 encourages believers to focus on whatever is true, noble, right, pure, lovely, and admirable. How does focusing on these things affect your mental health and spiritual well-being? What specific thought patterns may you need to "unlearn" in order to be more aligned with God's will and purpose for your life?

5. Second Corinthians 10:5 calls us to take every thought captive and make it obedient to Christ. What does this look like in practice?

6. How do we discern which thoughts need to be taken captive? What is a personal example where you had to actively take a thought captive and redirect it toward something more Christ-centered?

7. What role does self-awareness play in transforming your thought life, and how does the H.A.L.T. technique (hungry, angry, lonely, tired) help in this process?

CHAPTER 11
THE FREQUENCY OF PRAYER

SCOTT OATSVALL

Prayer is at the heart of the Christian life and is the cornerstone of faith. The frequency of prayer offers a pathway to communicate directly with God, express gratitude, seek guidance, and find a personal connection to the Creator of the universe. The New Testament speaks repeatedly of prayer, encouraging believers to engage in it often and consistently. One of the most striking commands is found in 1 Thessalonians 5:17, where Paul exhorts, "Pray without ceasing" (ESV). Yet, how often should we pray? Is prayer a once-daily discipline, or does it need to be a continuous thread woven through every aspect of our lives? In this chapter, we will explore the frequency of prayer, looking at biblical teachings, historical practices, and personal testimonies that highlight the importance of consistent, frequent communion with God.

In the realms of both the seen and unseen, the natural and spiritual worlds, frequency plays a pivotal role in influencing how things resonate and align with one another.

Every sound, from the hum of a refrigerator to the delicate song of a bird, exists within a specific frequency. Similarly, thoughts and prayers are not merely abstract concepts as we discussed in previous chapters; they are forms of energy that vibrate at their own frequency. So let's take a closer look as we explore the frequency of prayer—how it functions, its significance in both modern and biblical contexts, and how it shapes the world around us.

THE SCIENCE OF SOUND AND FREQUENCY

Sound is a physical phenomenon. It is a vibration that travels through air (or other mediums) and is interpreted by our ears. The human range of hearing spans from about 20 Hz to 20,000 Hz, though these numbers vary from person to person. Every object, being, or phenomenon has a specific vibrational frequency, and sound is simply a reflection of those vibrations. In the world of science, sound frequencies are understood to influence everything from the materials we interact with to the emotional and physiological responses we experience.[1]

When we think of prayer, it is often seen as a deeply spiritual practice. Yet, in many ways, prayer—like sound—can be understood as a form of frequency. Prayer is not only about communicating with God but also about aligning oneself with the will of the Father and opening up a channel for that energy to flow. It's a vibration in both thought and voice, transcending physical borders and reaching into the spiritual plane. The act of praying raises one's internal frequency, which can affect the environment around them, their own state of being, and the larger spiritual atmosphere.

MODERN-DAY EXAMPLES OF THE FREQUENCY OF PRAYER

In contemporary society, the power of prayer has often been examined in psychological and scientific terms. One clear example of prayer's frequency influencing real-world outcomes can be found in the practice of "affirmative prayer" or "positive affirmation." These are deliberate, focused prayers or thoughts aimed at specific outcomes, designed to raise one's internal vibration.

For instance, in the world of health and wellness, there are numerous testimonies of people using prayer to heal or prevent illness. A common practice among many who believe in the power of positive prayer is "praying over" or "blessing" their bodies in moments of sickness. One example is the use of prayer in overcoming serious illnesses such as cancer. There are numerous accounts of individuals who, with the collective support of a community praying for their healing, saw remarkable recoveries against all odds. In these cases, the belief is that prayer activates a higher vibrational frequency and a healing frequency, which not only connects the individual with a spiritual force but also instigates a healing process, both spiritually and physically.

Dr. Mitchell Gaynor, an oncologist who worked with cancer patients, explored the connection between prayer and healing. He argued that the mind-body connection is powerful and that focusing on healing thoughts can alter the body's physiological state. Though his conclusions did not always point to a strictly spiritual understanding of prayer, he noted that those who engaged in practices like prayer or visualization tended to heal at higher rates than those who did not. This would suggest that prayer, as an energetic practice, impacts the body's frequency, thereby creating conditions conducive to healing.

Similarly, in the realm of relationships, prayer is often employed as a tool to improve communication, foster forgiveness, and strengthen connections. The frequency of prayer in these contexts often manifests as a deep sense of peace, emotional alignment, and mutual understanding between partners. You may have heard the phrase "A family that prays together, stays together." Studies have shown that when couples pray together, they often experience increased intimacy and resilience in their relationships. The act of praying with one's partner as well as family and friends creates a spiritual resonance, raising both individuals to a higher frequency of love and empathy, which can manifest in practical, tangible ways. My grandmother Evelyn Bright would always say, "A family that prays together stays together."

PRAYER AND FREQUENCY IN THE BIBLE

The Bible provides numerous examples of the transformative power of prayer. Many of these examples illustrate how prayer shifts not only the heart of the individual praying but also the environment in which they live. There is a profound biblical truth that aligns with the concept of frequency: when we pray, we attune ourselves to God's will, thus opening up supernatural channels that affect not just our internal state but the world around us.

In the book of James, the apostle writes, "The prayer of a righteous person is powerful and effective" (5:16). The frequency of a prayer, according to James, is directly tied to the righteousness of the one offering it. In other words, when one's life is aligned with God's will—living righteously—the frequency of their prayer carries greater power. Righteousness, in this sense, could be understood as a state of being in tune with divine frequency. A righteous life aligns with God's purposes, and prayer emanates with a spiritual vibration that is aligned with His will.

One of the most notable prayers in the Bible is the prayer of Elijah in 1 Kings 18:36–37, where he prayed for fire to come down from heaven to prove God's sovereignty. His prayer, which was succinct yet fervent, was answered immediately with a supernatural occurrence that demonstrated the potency of prayer. The frequency of Elijah's prayer was a direct reflection of his faith, obedience, and divine alignment. His prayer carried power because it resonated with God's will, showing that frequency is not just about the words spoken but about the resonance between the human spirit and God's will.

Another profound example is the prayer of Hannah in 1 Samuel 1:10–20. Hannah, who was barren and deeply troubled, prayed earnestly to the Lord for a son. The Bible tells us that "she was in bitterness of soul" (v. 10 NKJV) and she wept in prayer. Her prayer was heartfelt and intense, and it came from a place of deep personal need. In response to her prayer, the Lord granted her request, and she bore a son, Samuel.

Hannah's prayer is a demonstration of how sincere, fervent prayer can shift the spiritual atmosphere. The frequency of her prayer, coming from a place of desperation and genuine faith, attracted divine intervention and produced tangible results.

THE ROLE OF FASTING IN INCREASING THE FREQUENCY OF PRAYER

In both the Bible and modern practice, fasting is often associated with prayer. Fasting is a means of intensifying one's connection to God, clearing the mental and physical space to increase the frequency of prayer. In Matthew 17:21, Jesus spoke of how certain spiritual victories—such as casting out powerful demons—required both prayer and fasting. Fasting, when paired with prayer, helps to elevate the spiritual frequency by removing distractions and focusing a frequency on the mind and heart of God.

In modern-day applications, many people fast as a way to break through spiritual stagnation. The act of removing food or other distractions for a set period is seen as a way to attune oneself to a higher frequency of divine energy. By abstaining from physical needs, the individual increases their spiritual awareness and draws closer to God's will.

CONCLUSION

The frequency of prayer is not just a metaphor—it is a dynamic, energetic force that aligns the individual with supernatural and divine power. Whether in the modern-day practice of healing, relationships, or achieving specific goals, prayer's frequency has a tangible, measurable impact on the individual and their environment. Biblical examples further underscore the importance of aligning one's life with God's will, ensuring that the frequency of prayer resonates with divine purposes. By understanding prayer as a frequency, we see that it is not only a tool for communication with God but a powerful means of transformation, capable of changing both the heart and the physical world around us.

PERSONAL REFLECTION

1. What does it mean to think of prayer as a frequency? How does this idea change your perception of how prayer works?

2. How can prayer be integrated into daily life beyond designated times of prayer (e.g., during work, in the midst of relationships, when facing challenges)?

3. What are some practical ways you can align your life more closely with God's will, to ensure that your prayers resonate with His frequency?

4. In both the story of Elijah's prayer and Hannah's prayer, we see the power of fervent, passionate prayer. How does the intensity of our prayer affect the spiritual frequency and its outcomes?

5. Reflect on a time when you prayed with intense faith or desperation. How did that prayer shape your heart or circumstances? What did you learn from that experience?

6. How can we incorporate prayer into our daily lives, especially in areas of physical or emotional healing, to increase our own vibrational frequency?

7. It is well documented that couples who pray together often experience greater intimacy and resilience. How can the frequency of prayer strengthen relationships?

CHAPTER 12
THE FREQUENCY OF METALS

SCOTT OATSVALL

Throughout history, metals have been revered not only for their material value but also for their mysterious, potent, and healing properties. In the ancient world, metals were often believed to possess spiritual and healing frequencies that could influence both the physical and metaphysical realms, the seen and unseen realms.

As we have discovered in the previous chapters, frequencies have gained renewed attention, as scientists and researchers explore how different metals resonate at distinct frequencies and how these vibrations can affect the human body. From gold's resonance at 5.11 MHz to the effects of lead and copper, the ancient understanding of metals and their frequencies is not so far removed from the emerging science of bioenergetics and frequency healing used as therapies today. In this chapter, we will explore the healing properties of metal. This chapter will look at the benefits and some of the risks that these amazing materials have on both the physical and metaphysical.

THE FREQUENCIES OF METALS: A SACRED GEOMETRY

The idea that metals have specific frequencies is grounded in the notion that everything in the universe, including the human body, operates on a vibrational level. In other words, if something has matter, it has space. If it has space, it has energy. If it has energy, it has frequency. And if it has frequency, it can be measured. Ancient cultures, including the Egyptians, Greeks, and Hebrews, were aware of this connection, although they may not have had the technical language to express it in the way we do today.

Metals, due to their inherent crystalline structures, vibrate at certain frequencies that can influence their surroundings. The resonance frequency of a metal is essentially the natural rate at which it oscillates when energized, and this frequency can affect the energy fields of the human body. This notion ties into both the ancient and modern understanding of the interconnectedness of all things—mind, body, and spirit. Let's take a closer look.

GOLD (5.11 MHZ)

Gold, often called "the metal of the gods," has been cherished for millennia for its purity, beauty, and inherent ability to conduct energy. The resonance frequency of gold is typically measured at 5.11 MHz, a number that may seem arbitrary, but it holds profound significance in both science and the realm of spiritual symbolism.

In biblical tradition, gold is frequently mentioned as a symbol of divine glory, authority, and perfection. The most famous example of this is the Ark of the Covenant, which was constructed with gold to house the Ten Commandments (Exodus 25:10–22). Gold's association with purity and divine presence is not only symbolic but may also be related to its unique vibrational properties.

In the human body, gold is believed to have a harmonizing effect, promoting balance and restoring equilibrium to energy fields. In holistic medicine, particularly in practices like acupuncture and bioenergetic healing, gold is used to stimulate the body's electromagnetic fields. Some proponents of vibrational medicine believe that gold's high-frequency vibration helps the body overcome emotional and spiritual blockages, allowing for a deeper connection to God.

The ancient Egyptians were among the first to recognize the sacred properties of gold. Pharaohs were buried with gold artifacts, believing that these objects would help them connect to the gods in the afterlife. The gold they used in their tombs was not merely for ornamental purposes but was thought to carry a divine energy that would transcend the mortal realm. This belief is mirrored by the modern understanding of gold's ability to amplify and channel subtle energy fields, promoting healing and vitality.

SILVER (1.07 MHZ)

Silver, another metal of significant spiritual and medicinal value, resonates at a frequency of approximately 1.07 MHz. Historically, silver has been associated with the moon, reflecting its qualities of intuition, protection, and reflection. In ancient times, silver was thought to have protective properties, guarding against negative energies and physical illnesses.

The Bible also refers to silver as a symbol of purity and value. Proverbs 10:20 states, “The tongue of the righteous is choice silver,” suggesting that silver represents righteousness and truth. This aligns with the idea that silver’s frequency helps to clear and purify the body’s energy field, much like it was believed to purify water and cleanse wounds in antiquity.

Silver is often used in modern energy healing practices to enhance mental clarity and emotional stability. Some research suggests that silver, through its vibrational frequency, can also promote the regeneration of tissue and cells, making it a powerful tool for healing chronic illnesses and injuries. Just as silver was considered a “cleansing” metal in ancient cultures, its resonance today is thought to aid in detoxification on both the physical and energetic levels.

COPPER (21.1 MHZ)

Copper resonates at a much higher frequency than gold and silver, measuring around 21.1 MHz. Ancient cultures, particularly the Egyptians and the Sumerians, used copper extensively in their healing practices. Copper was used to treat various ailments, from wounds to infections, and its resonance was thought to enhance the body’s natural ability to heal itself.

In the Bible, copper is mentioned in relation to the construction of the tabernacle and Solomon’s temple, where it was used for various tools, utensils, and pillars (1 Kings

7:13–14). Copper's ability to conduct both heat and electricity gave it practical applications in the ancient world, but its spiritual significance was also profound. In the context of biblical stories, copper is often associated with strength, protection, and restoration.

The vibrational frequency of copper is believed to help open pathways for the flow of energy within the body, particularly by increasing the efficiency of the circulatory and nervous systems. Its high frequency makes copper particularly effective in removing blockages and facilitating the flow of blood throughout the body. Ancient practitioners of alchemy and early forms of energy medicine recognized that copper could amplify the healing properties of other metals, enhancing their effects.

IRON (1.1 MHZ)

Iron, a metal that has shaped human civilization for millennia, resonates at approximately 1.1 MHz. In the Bible, iron is frequently associated with strength and resilience. In Deuteronomy 3:11, for example, the iron bed of King Og is described as a symbol of his great power. Iron's symbolic significance is also reflected in its practical applications, from weapons and tools to its role in modern-day technology.

While iron's frequency is relatively low compared to that of gold or copper, it still plays an essential role in the body's energetic and physical health. Iron is crucial for the production of hemoglobin in red blood cells, which carry oxygen throughout the body. Its vibrational properties are thought to help the body absorb and distribute vital nutrients, enhancing overall vitality and endurance.

Iron's grounding energy may also help stabilize emotional and physical imbalances. In modern vibrational medicine, iron is used to help individuals reconnect with their sense of purpose and inner strength. Just as iron has been integral in forging tools and weapons throughout history, its frequency is believed to help forge resilience in the human spirit.

LEAD AND MERCURY: DISRUPTORS OF FREQUENCY

On the darker side of metals, lead and mercury are frequently cited as toxins that disrupt the body's natural frequencies. Both metals have historically been used in various applications, from vaccines and medicine to art, but their detrimental effects on health are now well-known.

Lead, with its low frequency, has been linked to cognitive impairment, particularly in children. In ancient times, lead was used in cosmetics, plumbing, and even in food preparation, with disastrous effects on health. Its resonance interferes with the body's ability to maintain balanced energy, contributing to chronic illness and fatigue.

Similarly, mercury, once thought to possess alchemical properties capable of transforming base metals into gold, is a toxic disruptor of the body's frequencies. Even in small amounts, mercury can cause neurological damage and disrupt the body's energetic equilibrium and immune system. In alchemy, mercury was symbolized as both a purifier and a poison, and its dual nature reflects its ability to destabilize the body's energy field.

CONCLUSION

Metals have always held a place of reverence in human history. From their use in sacred rituals to their role in promoting healing, these elements have been seen as more than just physical substances. They are frequencies in tune with the rhythms of how God created the universe. Gold, silver, copper, iron, and even lead and mercury all carry vibrations that can influence the human body in profound ways. Ancient civilizations understood this, as evidenced in their myths, religious texts, and healing practices.

As modern science and medicine continues to uncover the mysteries of bioenergetics, the ancient wisdom of metal frequencies finds new relevance. The biblical stories of gold and silver, the use of copper in ancient temples, and the reverence of iron as a symbol of strength and resilience echo the understanding that metals carry frequencies far beyond their physical properties.

By gaining a greater understanding of these frequencies, humans can tap into an ancient knowledge of balance, healing, and connection to the one true God!

PERSONAL REFLECTION

1. The Bible consistently warns against the dangers of idol worship and the consequences of clinging to worldly treasures (see Matthew 6:19–21). Just as lead and mercury disrupt the body's natural functioning, idolatry disrupts our relationship with God. How do we allow "toxic" influences to affect our spiritual lives today? What are the "idols" or distractions in your life that might be corrupting your spiritual health?

2. Consider these:

- **Gold in Your Life:** Gold symbolizes purity and divine presence. How do you invite God's purity and presence into your life? Are there areas where you need God's holiness to be more evident in your thoughts, actions, or relationships?

- **The Purifying Power of Silver:** Proverbs speaks of silver as representing purity. In what areas of your life do you feel the need for purification or refining? How can you seek God's purification in those areas?

- **Strength from Iron:** Iron is a symbol of strength in Scripture. Reflecting on King Og's iron bed (Deuteronomy 3:11), what areas of your life do you need to call on God for greater strength? How can God's strength be made perfect in your weakness (see 2 Corinthians 12:9)?

- **Copper and Healing:** Copper is said to resonate at a frequency that aids healing. While we may not fully understand how this works, in what ways can we invite God's healing presence into our physical, emotional, and spiritual lives?

- **Toxic Metals—Lead and Mercury:** Lead and mercury are toxic and disruptive. What are the "toxic" influences in your life—whether physical, emotional, or spiritual—that prevent you from living in harmony with God? How can you detoxify your life spiritually, emotionally, and physically to better align with God's will?

- **God's Presence in the New Jerusalem:** Revelation 21 describes the New Jerusalem as being filled with gold, precious stones, and other beautiful materials. How does this imagery encourage you about the eternal home God has prepared for you? What does it tell you about the beauty and perfection of God's kingdom?

PRAYER

Heavenly Father, we thank You for the lessons we have learned from Your Word about the significance of frequencies and their deep spiritual meanings. We ask that You help us invite Your presence into our lives like pure gold, purify us like silver, strengthen us like iron, and heal us like copper. Lord, we also pray that You help us identify and rid ourselves of any "toxic" frequencies that might be disrupting our heavenly frequencies and our spiritual walk with you. May we live in harmony with Your will, and may we always line up with the frequency of faith. And may our faith always be reflecting the purity, strength, and glory of Your kingdom. In Jesus's name, amen.

CHAPTER 13
SOUND, FREQUENCIES, AND VIBRATION

JOHN TUSSEY

Disclaimer: The information shared about my music in this book has not been evaluated by the FDA. The information, music, and CDs mentioned are not meant for diagnosing, treating, or curing illnesses or diseases of any kind. The experiences provided in this book are shared by real people who have had real experiences. These are not intended to be nor are they health claims.

"God said . . ." God used the sound, the frequencies, the vibrations, of His mighty, matchless, all-powerful voice to create. Wow! The Scriptures don't say, "And God imagined . . . And God pointed His finger . . . And God thought . . ." It is clear and evident that God used His voice to create. Everything in existence has sound waves in it because of how God created. Whether living or not, everything on earth and in the universe is vibrating with its own signature sound. All the cells and organs in our bodies are constantly vibrating. We are a walking symphony of sound that God can hear.

In the last fifteen years, interest in "frequencies" has increased. A frequency is the number of times per second a sound wave passes a fixed point. Frequencies are measured in hertz, abbreviated Hz. So, if there are sixty sound waves per second passing a fixed point, that is 60 Hz. To make it simpler, we can say that frequencies are measured sound waves. As you can imagine, there are countless frequencies.

Whenever I hear someone talk about the frequency or frequencies of heaven, as if there is one or a particular set of frequencies that is spiritually exclusive, that shows me that some believers are swallowing erroneous information and are not thinking about or researching what they're being told or taught about this subject. Can you imagine how many frequencies there are in heaven? I think you get my point.

Just as our bodies, as well as all of creation, are vibrating with many frequencies blending together, the music we listen to is a combination of many frequencies blended together. I have been asked a number of times how to tell if music is good or bad. I believe that Holy Spirit has given me a simple explanation for that.

There are a few things to consider. First, what was the intent of the one or ones recording the music? Then, what kind of spirit was the person, group, choir, and/or orchestra moving with and motivated by while they were singing and/or playing the music they recorded? If there was an evil spirit involved, every time that music is played back, an evil spirit will be attracted to that music. Those who are sensitive in the spirit can pick that up within a few seconds. If there was no evil present and those singing and/or playing were doing so out of their emotions, which is the case for many pop songs, then you most likely won't feel any bad vibes from that music, as long as they were singing about positive things, because there was no ill intent. Many songs are like this. If Holy Spirit was anointing those recording the music, He will show up live every time that music is played back. Music is a language of the heart and can be an expression of the soul or the spirit or both.

The importance of which frequencies are used in music is something I have been researching and engaging with since March 2011. To give some background, in 2008 I started listening to teaching CDs by a man named David Van Koevering who lived in Cleveland, Tennessee.

David was a quantum physicist, an inventor (he invented the technology that preceded our CDs and DVDs —writing data on film), a musicologist, and a visionary. He worked with Robert Moog, who invented the Moog synthesizer, over a period of approximately forty years. David created the market for the Moog synthesizer, taking it across the United States and into twenty-six countries. He also contributed to its early development, and that instrument forever changed the music industry. In 1980, David also invented a keyboard instrument called the Orchestron, and in 2000, with a team of 124 employees, he developed the Van Koevering Touch-Screen Interactive Piano, which Bill Gates called the most sophisticated use of Windows he had seen up to that time. This instrument has 636 patents on it, so David obviously holds a significant place in the history of music technology.

In his captivating teachings, more than once David talked about an instrument that produced the frequencies of the Periodic Table of Elements (PTOE). This immediately caught my attention, and the desire was planted in my heart to use the technology that enabled this. David's teachings shared fascinating things about the realm of quantum physics—things I had never heard before. I listened intently to his CDs for seven months, being intrigued by his teachings. Later that year, I spoke to his wife Becky on the phone, and after getting her approval, I mailed two of my keyboard instrumental CDs, "Serenity" and "Flowing," to David.

On May 11, 2009, at 6:09 a.m., David called me and shared some extremely encouraging words about the CDs I had sent him. He told me that his music room contained fourteen one-foot stacks of vinyl albums and that my "Serenity" CD was beyond any of them. In June 2009, while here on Oahu, he told me that my "Flowing" CD was his favorite piano/synthesizer recording of all time and gave me a handwritten endorsement for it. I was highly honored, as David had worked with keyboard legends like Keith Emerson, Rick Wakeman, and Stevie Wonder, along with many other well-known musicians from the 1960s,'70s, and '80s.

During our phone conversation on May 11, David also shared that he wanted to overlay frequencies of the PTOE onto the "Serenity" CD for the purpose of healing. Wow! This was more than I originally desired! I'm now actually able to play and record with the technology that I heard David speak about while listening to his teaching CDs in

2008! And I'm going to be recording these CDs for the purpose of healing, which is near and dear to my heart.

David was a strong Christian as I am, with both of us having belief systems strongly rooted in God's Word. The Scriptures teach that healing is for today—for spirit, soul, and body. I believe that because of Jesus's stripes and His finished work on the cross, He is THE source of healing. I also believe He has different ways of getting healing to us. This is how my exciting journey with the frequencies of the PTOE began.

When David and his wife Becky came to Oahu in June, July, and October 2009, he mentored me on how to record with the PTOE frequencies—to have them embedded in music with long sustained tones. We also spoke of the possibilities for where the CDs would be sold and how far they would reach into the world. David was convinced that amazing things would happen as I played the albums along with using the plasma projectors that he gave me to enhance the effectiveness of the frequencies in the music.

In March 2011, I learned how to use the technology to record with the PTOE frequencies. The inventor of the technology, who was a friend of David's, told me he had consulted with physicists to ensure the PTOE frequencies he calculated were accurate. I'm confident these are accurate frequencies of the PTOE because of the many testimonies I've received from those who have listened to my albums.

Around three years after my mentoring time with David Van Koevering, I was listening to one of his teachings from 2010. He told a story that has not only amazed me but also brought the realm of what my music carries into another level. In this teaching, David shared that his heart had shut off twelve times over a period of perhaps a year. When his heart shut off the twelfth time, he passed over and heard sounds, saw colors, and his bandwidth, as he called it, really went up. Some time after this experience, he attended an event in Nashville, Tennessee, where the instrument that played the PTOE frequencies was being demonstrated. David was one of several people who assisted in making that unique instrument, and the technology used for the PTOE frequencies in that instrument is the same technology I've used since 2011 to record with those frequencies.

After hearing the instrument with the PTOE frequencies during that event, David said, "We now have the Periodic Table of Elements that is the voice of God that I heard as the music on the other side." Referring to the instrument playing those sounds, David said, "All I could do was weep because I'd heard the sounds that night that I'd heard in heaven."

When I discovered I was recording with frequencies that David Van Koevering heard in heaven, the revelation was awe-inspiring and caused me to begin realizing how powerful those sounds were. Suddenly, I understood why David was so confident and absolutely sure that people would be healed when they experienced these PTOE frequencies in my music while also using the plasma projectors he gave me to enhance the effectiveness of the frequencies.

David had not revealed this story about hearing these sounds in heaven during the time he mentored me. I'm not sure why, but what an awesome discovery it was to hear him share this in that message in 2010, after I had already recorded one or two albums with PTOE frequencies.

Do I believe these frequencies are important, worth recording with, and accurately transposed into the human hearing range? A HUGE resounding YES!

Since then, I have recorded with thirty-seven frequencies of the PTOE found in the human body for the purpose of promoting health and wellness. These albums containing PTOE frequencies all *begin* with the word "Frequencies." Here are some testimonies from listeners:

> "After listening nightly, while sleeping, to *Frequencies of Thought*, my potassium level went from borderline low to halfway range. Now I don't have to take potassium tablets. I just listen to *Frequencies of Thought*!"
>
> Mahalo nui loa,
> Jackie O.
> Kapolei, Hawaii

> "I played John's music to my garden last year, and this year we have monster-sized tomato plants and squash that is coming out our ears."
>
> Roslyn K.
> Paradise, California

> "I played your music (*Frequencies of Silver and Gold*) in a meeting recently and everyone with pain/aching got healed."
>
> Nelson
> Texas

"When I started using the *Frequencies of Thought* and *Frequencies of the Heart* albums, I was walking with a stick and having chiropractic treatment for degenerative disc disease. At that time I had no reflexes in my left leg and had a trapped nerve also with nerve damage in my spine. Since listening to both these albums, my pain (which I was taking two types of morphine for) has decreased by 60 percent and I am no longer using a walking stick.

I attribute the reduction in pain to literally bathing myself in your albums, playing them all night and all day. Praise the Lord. I'm hoping to go back to work soon, as I've been off work for six weeks. This is amazing, I have such less pain and I can bend like I used to! Thank you so much for your healing albums!"

ADE T.

UNITED KINGDOM

"Hi, John! No pain in my back at all! I've now come off my Tramadol, which is a major achievement. I'd been on it ten (yes, ten) years! Now got a reduction plan with my doctor to come off short-acting and long-acting morphine! Hallelujah! I play your music every day in a psychiatric ward!"

ADE T.

UNITED KINGDOM

"I wanted to thank you for your music. I was at your presentation last week with an aching, clicking left jaw that had been steadily getting worse. It eventually started on the right also, so I was pretty concerned. I had the opportunity to listen to your music with the discs that amplify the frequencies (*Frequencies of Life* CD).

After less than ten minutes, my jaw completely went back to normal! No pain, no clicking. At home I used *Frequencies of Creation* during my prayer/meditation time and my nagging upper back pain also dissolved. I often hear of this kind of thing happening to other people, so I'm grateful I could experience it for myself with your healing music. Thank you so much for developing and sharing your gift, John!"

Denise H.
Honolulu, Hawaii

"I have had muscular/nerve pain in my legs, which immediately improved from the first time I listened to your *Frequencies of DNA* CD while falling to sleep, and it continues to improve. I have also been in the healing process of adrenal fatigue, which sent me to the hospital a little over a year ago, and I am seeing real progress there as well. Good, deep sleep is really important to healing the adrenals, and I find God is blessing me each night with this (especially when I listen to your frequency music).

I have more energy, which is huge as we have five children and I homeschool them all. This music

God has given you is huge! I pray people everywhere begin to take notice of the power in it to bless us spiritually, physically, and emotionally."

Shelly E.
Ocala, Florida

"*Frequencies of Abundance* is my most favorite CD. I listen to it almost every night. It helps me sleep soundly. I've noticed that each time I play this CD while resting, I discover new sound. This music truly does emanate from the eternal realm! I also would like to share a testimony regarding a friend of mine who experiences panic disorder. She specifically sought my help while experiencing an episode. During this time I began to play your *Frequencies of Abundance* CD with the personal knowledge that this anointed music would fill her with a certain peace. And it did just that! Her panic immediately turned into peace!"

Donna N.
Lansing, Michigan

"This music is a true gift from our amazing God!

I have felt incredibly good while listening to this music—light as a feather and relaxed, yet also energized by it. It seems as though every cell in my body is being rejuvenated. I have slept with the music playing and woken up with my entire body vibrating with its healing energies.

There is a tangible peace in the atmosphere whenever it plays. *Solfeggio Frequencies 2* is my favorite of all the 'Frequency' CDs John has made. I also love those with the Periodic Table of Elements, which are absolutely genius and delightful to listen to. There is healing in this music and I'm a believer in it! Everyone needs to get this music to see for themselves how wonderful it can make them feel!"

LORI T.

LAKESIDE, ARIZONA

"I recently received John Tussey's CD, *Frequencies Of Abundance*. I must share with you, I have known about frequencies for a long time, however John has brought it to us in a most extraordinary way, through his amazing God-given gift of music. He has recorded the music with the frequencies of the Periodic Table of Elements.

I have never heard anything like it! As I listen, every cell—yes, every cell in my body—responds to the frequencies! I experience extreme peacefulness. Any tension in my body literally melts away! Knee pain has ceased. Inflammation in my joints is healing. And as I continue to listen to these healing frequencies, I believe I will have more testimonies!

I believe God is using John's music to bless and heal people. And as a side note, I have noticed the plants I have sitting in the room where I listen to this music are also responding! I have seen the leaves gently move, and they are thriving! All of creation responds to frequency. I highly recommend obtaining John Tussey's CD, *Frequencies of Abundance*. Sit, lie, walk . . . listen.

You will be blessed."
JEANETTE S.
FOREST, VIRGINIA

I have also recorded several albums with the Solfeggio frequencies—396 Hz, 417 Hz, 528 Hz, 639 Hz, 741 Hz, 852 Hz. Much of the information about these frequencies found on the internet is either unprovable or false. However, I do believe in the validity of these frequencies and found some very inspiring, scientifically documented information about 528 Hz on the Net.

In 2010, John Hutchinson and his friend Nancy went down to Biloxi, Mississippi, after the disastrous BP oil spill. They took 6,000 pounds of audio equipment with them and brought it close to the water. They played 528 Hz and other frequencies toward the gulf for a twenty-four-hour period of time, and cleared oil in a one-mile radius from where they were set up. The scientific document that I have, showed the water had 7 ppm (parts per million) of oil before the frequency treatment and less than 1 ppm after the treatment. Do I believe the Solfeggio frequencies are worth recording with? Yes, I do, especially 528 Hz, because of the scientific documentation substantiating its effectiveness in the treatment of oil-polluted water, as well as testimonies I have received from listeners.

"This *Solfeggio Frequencies* CD is like none other. . . . There are no words in it, it is spiritual, it will touch your spirit, free you, and it will move your spirit into worship effortlessly. Can't explain it any other way! . . . It isn't just listening to music, IT

IS experiencing music in your body and spirit. Never have I experienced what I do when I listen to this music straight from heaven!"

MALINDA M.,
MICCO, FLORIDA

I have received frequency downloads from heaven over the last several years. In 2019, before going to bed one evening, I received five frequencies: 171.25, 271.25, 371.25, 471.25, and 571.25. One of the things that came to me about these frequencies is that new neural pathways would be opened up. The most recent frequency download came in May 2022. I was visiting family and had gotten COVID-19. I received this unexpected download one afternoon: 4961.53. I typed the number into the frequency generator in my phone, turned it on, and placed the phone on my chest. I immediately felt somewhat better. In July 2024, I released my thirtieth album, *Waves of Abundance*, and embedded 4961.53 in every track on the album, as I believe the Lord directed me to do. After the album release, I gave several copies away. Among those who received the new album was a minister friend, R. G., in Hawaii whom I have known for years. After listening to the album for the first time, she sent me this testimony:

"I played it last night . . . all night and I woke up healed from bronchial and nasal blockages. I bless the Lord who downloaded the frequencies to you to share. When it's the Lord, it's indescribable. He is Lord over all."

These are two prophetic downloads I received regarding the music I have been entrusted with, and explain what I call "Frequency Cellular Infusion."

Word received on January 26, 2014, at Kokokahi YWCA Atherton Hall between 7:09 and 7:18 p.m.: "The music you release cleanses the DNA at the sub-atomic level. That is the reason God released you to record and release it into the earth. Many will not connect with this, but on the other hand, many will. DNA will be released into new levels, even as Adam experienced right after creation, as this music enters the cells. As the music goes forth into the cells, the cells will respond with joy, recognizing the vibrations as those first released from heaven. When David Van Koevering heard the sounds in heaven, you heard in his testimony that he recognized them as the Periodic Table of Elements frequencies. These frequencies released into your music as PTOE frequencies in the audible spectrum are from heaven."

I gratefully received this word from the Lord on February 3, 2014: "There is a grateful receiving of these PTOE frequencies by the cells in the body, because the body recognizes these as frequencies of life—oxygen, carbon, hydrogen, nitrogen, etc. All vitamins and minerals are made from the elements in the Periodic Table. All cells are frequencies at the sub-atomic level. These frequencies of the PTOE are like taking nutritional supplements in frequency form, but not a substitute for taking supplements. We still need them in physical form, but these frequencies of the PTOE go into the cells in a stream, flowing like the Gulf Stream, and are taken into the cells. The cells absorb them with joy and gladness, recognizing them as refreshing, renewing, energizing, stimulating, cleansing life elements that are already resident within the cells. The cells welcome these frequencies of life as dearly beloved family members are welcomed and embraced. The blood welcomes these frequencies as life-giving elements that cleanse and renew."

Again I would ask, are there frequencies that bring healing? Frequencies that could be considered important to record with? Based on the testimonies you have read, I have to say yes.

Somcthing interesting . . . By placing a glass of water, a plate of food, or a piece of clothing near or on the speaker where this music is playing from, your water, food, or clothing will be infused with whatever frequency or frequencies are embedded in the music.

As you research frequencies, please remember that you cannot believe everything you read on the internet. There is a lot of misleading, unprovable, and false information about frequencies there. Always look for information that is scientifically and/or medically proven.

CHAPTER 14

SALVATION: THE MOST POWERFUL GIFT OF ALL

JOAN HUNTER

While we have written about the power of the frequencies of heaven and the many positive effects they can have, welcoming Jesus as your Lord and Savior produces the ultimate change anyone can experience. Welcoming God's Son into an open and willing heart is truly receiving the greatest gift of all. Despite the many ways each of us has fallen into sin, God still loves us. Jesus spoke these words: "For God so loved the world, that He gave His only begotten Son, that whoever believes in Him shall not perish, but have eternal life" (John 3:16).

The promised Holy Spirit will never leave us or forsake us. No matter how bad we have been, He still loves us because we are His creation, and He longs for us to become His children through faith in Jesus. God first loved you and demonstrated that by sacrificing His only Son for your sins thousands of years before you were born (see Romans 5:8).

John 3:16 contains the essential elements you need to know the truth about God:

- God's extravagant, self-sacrificing love . . .
- compelled Him to send a gift to earth to save it —His Son, Jesus.
- If any choose to believe in Jesus and receive God's precious gift,
- they would no longer be condemned to eternal death separated from God,
- but enjoy life after death, experiencing the pleasures of God forever.

The rest of the Bible reinforces this one message from Genesis to Revelation. The book of Romans, written by the apostle Paul, has a number of verses that outline what salvation is and how to receive it. Read these verses and the truths they declare.

EVERYONE NEEDS SALVATION

- **Romans 3:23:** "For all have sinned, and come short of the glory of God." Our sin has separated us from God. We've come up short and need a Savior to help bridge the gap.

- **Romans 6:23:** "For the wages of sin is death; but the gift of God is eternal life through Jesus Christ our Lord." The penalty for our sins is death, and eternal death separated from God and anything good is our destination without Jesus. But God has offered us another option: eternal life *through Jesus* if we will accept it.

- **Romans 10:9:** "If you confess with your mouth Jesus as Lord, and believe in your heart that God raised Him from the dead, you will be saved." We receive the gift of salvation and enter into a relationship with God by doing what Romans 10:9 instructions: confession and belief.

- **Romans 10:13:** "For everyone who calls on the name of the Lord will be saved." The Bible promises that God answers those who call out to Him with salvation. Anyone who trusts in Jesus Christ and calls on His name as Lord and Savior will be saved. There is no age limit and no qualifying criteria. Anyone at anytime, anywhere can call on His name and be saved. Even with a final breath before death, God hears that request.

THE BENEFITS OF SALVATION

- **Romans 5:1:** "Therefore, since we have been justified through faith, we have peace with God through our Lord Jesus Christ." Our faith in Christ makes us right with God and we enjoy peace.

- **Romans 8:1:** "Therefore, there is now no condemnation for those who are in Christ Jesus." Because Jesus died on the cross to pay for the sins of all mankind, if you believe, you will never be condemned for your sins. You will live eternally with God and now the Holy Spirit lives within you.

- **Romans 8:38–39:** "For I am convinced that neither death nor life, neither angels nor demons, neither the present nor the future, nor any powers, neither height nor depth, nor anything else in all creation, will be able to separate us from the love of God that is in Christ Jesus our Lord." Because of your faith in Christ, you will never be separated from God. His love is within you and is eternal.

QUESTIONS TO ASK YOURSELF

1. Do you admit that you are a sinner—that your moral failures have separated you from God and without receive the gift of salvation from Jesus, you are headed for eternity separated from God?

2. Do you believe that Jesus, God's Son, died for your sins on the cross?

3. Will you forsake your old life and follow Jesus and the life He has for you?

PRAYER OF SALVATION

If you want to receive Jesus Christ into your heart as your Lord and Savior, pray this prayer:

> Father, I know that I have sinned and my sins separate me from you. I turn away from these sins and ask you now to remove them from me. Thank you, Jesus, for the sacrifice you made on my behalf and that you have taken upon yourself what I deserve. I am truly sorry for what I have done in the past. Jesus, you are Lord and I believe that you died on the cross for my sins, resurrected from the dead, and are alive. Today, I turn away from my way of doing things, embrace all You have for me, and choose to follow you. Be both my Savior and King. I willingly and joyful submit to your leadership in my life. Holy Spirit, baptize me with fire and empower me to be a witness for Jesus's name. Guide me into all you have for me, in Jesus's name. Amen!

NOTES

2. UNDERSTANDING FREQUENCIES

1. *Merriam-Webster.com Dictionary*, s.v. "hertz," accessed January 3, 2025, https://www.merriam-webster.com/dictionary/hertz.
2. "The Frequency of Musical Notes," University of St. Andrews.
3. "Visible Light Spectrum," National Optical Astronomy Observatory.
4. "Radio Wave Frequencies," Federal Communications Commission (FCC).
5. "Pythagorean Harmonics," Encyclopedia Britannica.
6. "Heinrich Hertz's Experiments," Physics Today.
7. "Maxwell's Equations," NASA, accessed January 3, 2025, https://www.grc.nasa.gov/www/k-12/Numbers/Math/Mathematical_Thinking/maxwells_equations.htm.
8. "The Science of Pendulum Motion," Physics Education Technology.
9. "Resonance and Engineering," Journal of Applied Physics.
10. Marshall Brain and Talon Homer, "How Radio Works," HowStuffWorks, accessed January 3, 2025, https://electronics.howstuffworks.com/radio.htm.
11. "The Spectrum of Wireless Communication," IEEE Spectrum.
12. "Frequency and Pitch," Music Theory Online.
13. "Harmonics in Music," *Cambridge University Press.*
14. "Human Frequency Perception," *Journal of the Acoustical Society of America.*
15. "Color and Light Frequencies," *American Physical Society.*
16. "Ultrasound and MRI Imaging," *Medical Physics Journal.*
17. "Seismic Waves and Earthquake Prediction," *Seismological Society of America.*

3. THE SCIENCE OF VIBRATION

1. "Fundamentals of Harmonic Oscillators," *Physics Education Technology.*
2. "Resonance and Structural Failures," *Journal of Structural Engineering.*

3. "Frequency and Pitch in Music," *National Institute of Standards and Technology (NIST).*
4. "The Tacoma Narrows Bridge Collapse," *American Society of Civil Engineers.*
5. "Brainwave Frequencies and Their Implications," *Neuroscience and Biobehavioral Reviews.*
6. "The Resonance of Musical Instruments," *The Journal of the Acoustical Society of America.*
7. "The Physics of the Human Voice," *Voice Science and Technology.*
8. Avey H, Matheny KB, Robbins A, Jacobson TA. Health care providers' training, perceptions, and practices regarding stress and health outcomes. J Natl Med Assoc. 2003;95(9):833, 836-84514527051, https://jamanetwork.com/journals/jamainternalmedicine/fullarticle/1392494#google_vignette.
9. "Seismic Waves and Earthquake Analysis," *Seismological Research Letters.*
10. "Engineering Damping Systems in Skyscrapers," *International Journal of Engineering Research.*
11. "Resonance and Vibration Control in Vehicles," *Automotive Engineering Journal.*
12. "Frequency Modulation in Telecommunications," *IEEE Transactions on Communications.*
13. "Ultrasound Imaging and Frequency Use," *Medical Imaging Review.*
14. "Innovations in Materials Science and Vibration," *Advanced Materials Research.*

4. FREQUENCY IN THE HUMAN BODY

1. C. A. Czeisler and E. B. Klerman, "Circadian and sleep-dependent regulation of hormone release in humans," *Science*, 284(5422), (1999): 2177–2181.
2. E. Niedermeyer and F. L., *Electroencephalography: Basic Principles, Clinical Applications, and Related Fields* (5th ed.) (New York: Lippincott Williams & Wilkins, 2004), ___.
3. X. Gao, J. Li, Y. He, and S. Zhang, "Effects of binaural beats on relaxation: A systematic review," *International Journal of Clinical and Experimental Medicine*, 7(6), (2014): 1718–1725.
4. J. McKinney, J. "Vibroacoustic therapy: A review of its role in health and wellness." *Complementary Therapies in Clinical Practice*, 18(3), (2012): 134–139.
5. Y. Zeng, F. Wang, and S. Li, "The effects of Tibetan singing bowls on

mindfulness and relaxation," *Journal of Alternative and Complementary Medicine*, 21(10), (2015): 636–641.

6. M. S. Schwartz, and F. Andrasik, *Biofeedback: A Practitioner's Guide* (3rd ed.) (Greensboro, NC: The Guilford Press, 2003), ___.
7. L. Chaieb, T. P. Reber, and A. Lehmann, "Binaural beats and their effects on sleep and cognitive performance." *Psychophysiology*, 52(6), (2015): 710–717.
8. S. Zeki and M. E. Duffy, "Effects of vibroacoustic therapy on chronic pain management: A clinical case study," *Journal of Pain Management*, 3(2), (2010): 129–134.
9. R. T. Thibault and A. Raz, "Neurofeedback and the modulation of brain activity in PTSD and TBI: A review of current research," *NeuroRegulation*, 4(1), (2017): 29–42.
10. L. Goodwin and R. McLaughlin, "The integration of sound therapy with traditional medicine for pain and anxiety reduction," *Journal of Alternative and Complementary Medicine*, 23(7), (2017): 509–517.

5. FREQUENCIES AND EMOTIONS

1. E. Brattic and M. T. Pearce, "The neural correlates of music-evoked emotions," *Frontiers in Psychology*, 4, 138.
2. M. V. Thoma et al., "The impact of a music intervention on the treatment of depression," *Psychotherapy Research*, 23(4), (2013): 424-433.
3. J. LeDoux, "Emotion circuits in the brain," *Annual Review of Neuroscience*, 23(1), (2013): 155–184.
4. R. McCraty and F. Shaffer, "Coherence: Bridging Personal, Social, and Global Health," *Global Advances in Health and Medicine*, 4(2), (2015): 24-37.
5. M. Gunkel and J. Krawietz, "Sound therapy in the treatment of mental health issues: A systematic review," *Journal of Music Therapy*, 54(1), (2017): 1–27.
6. J. Bradt and C. Dileo, "Music interventions for mechanically ventilated patients," *Cochrane Database of Systematic Reviews*, 2014(12).
7. K. L. Padmanabhan, et al., "The effects of binaural beats on anxiety: A systematic review and meta-analysis," *Psychological Bulletin*, 144(6), (2018): 573–589.
8. A. Maratos, et al., "Music therapy in the management of psychiatric disorders: A systematic review." *Psychological Medicine*, 38(11), (2008): 1475–1485.
9. A. Khatri, et al., "Neuroimaging studies of music and emotion: A

systematic review," *Neuroscience & Biobehavioral Reviews*, 121, (2021): 97–110.

10. M. Cormack, et al., "Integrating music therapy and mindfulness for emotional regulation," *Frontiers in Psychology*, 11, (2020): 579–590.

6. FREQUENCIES IN COMMUNICATION

1. A. Smith, "The Fundamentals of Radio Broadcasting," *Journal of Media Technology*, vol. 12, no. 3, 2020, pp. 45–60.
2. J. Doe, "Television Technology and FM," *Broadcasting Studies Quarterly*, vol. 8, no. 2, (2021): 22–35.
3. R. Johnson, "Vocal Frequency and Emotional Communication," *Psychology of Sound*, vol. 15, no. 1, (2019): 10–25.

7. FREQUENCIES AND THE HOLY SCRIPTURES

1. "Music and Worship in the Psalms," Biblical Archaeology Review.
2. "The Fall of Jericho: A Biblical Account of Sound and Vibration," Journal of Biblical Literature.
3. "Kabbalistic Interpretations of Divine Frequencies," *Journal of Jewish Mysticism.*
4. "Divine Manifestations and Sound in Exodus," *Old Testament Studies Review.*
5. "The Still Small Voice in 1 Kings," *Biblical Interpretation Journal.*
6. "The Role of Chanting in Biblical Worship," *Journal of Religious Music.*
7. "David's Harp and Spiritual Healing," *Ancient Israel Studies.*
8. "Creation and Divine Speech in Genesis," *Journal of Biblical Creationism.*
9. "Cosmic Order and Divine Frequency in the Bible," *Biblical Cosmology Review.*
10. "Integrating Biblical Texts with Contemporary Frequency Concepts," *Journal of Religion and Science.*

8. THE NEW AGE MOVEMENT AND THE HIJACKING OF FREQUENCIES

1. "The Origins of the New Age Movement," Encyclopedia of Religion and Society.
2. "Chakras and Energy Fields," Journal of Alternative and Complementary Medicine.

3. "Healing Frequencies and Sound Therapy," International Journal of Sound and Vibration.
4. "The 432 Hz Controversy: Science and Myth," Music Theory Online.
5. "Crystal Healing: A Scientific Review," Journal of Evidence-Based Complementary & Alternative Medicine.
6. "Commercialization of Alternative Health Products," Journal of Health Economics.
7. "Pseudoscience and Scientific Terminology in New Age Practices," Science and Education Journal.
8. "The 528 Hz Frequency and Health Claims," Journal of Sound and Vibration.

9. USING FREQUENCIES IN ALTERNATIVE MEDICINE

1. "Principles of Frequency in Alternative Medicine," *Journal of Complementary Therapies.*
2. "Frequency-Specific Microcurrent Therapy: Mechanisms and Applications," *Alternative Medicine Review.*
3. "Bioresonance Therapy: An Overview," *Journal of Alternative and Complementary Medicine.*
4. "Sound Therapy and Vibrational Healing," *International Journal of Sound and Vibration.*
5. "Tuning Fork Therapy: A Review," *Complementary Therapies in Medicine.*
6. "Binaural Beats and Cognitive Enhancement," *Psychology of Music.*
7. "Solfeggio Frequencies and Their Healing Properties," *Journal of Sound Healing.*
8. "Pulsed Electromagnetic Field Therapy: Clinical Applications," *Journal of Biomedical Science.*
9. "Transcranial Magnetic Stimulation for Mental Health," *Neuropsychiatric Disease and Treatment.*
10. "Evaluating Frequency-Based Therapies: Evidence and Challenges," *Journal of Integrative Medicine.*

10. THE FREQUENCY OF THOUGHTS

1. W. J. W. McCarty, "Quantum Field Theory," *Physics Review Letters*, vol. 110, no. 2, (2013).
2. T. P. Markham et al., "Neuroplasticity and Brain Development," *Journal of Neuroscience*, vol. 32, no. 12, (2012).

3. M. G. Smith, "The Effect of Learning New Skills on Brain Plasticity," *Neuroscience & Biobehavioral Reviews*, vol. 34, no. 5, (2010).
4. J. K. Allen et al., "Understanding Motivation in Behavior Change," Health Psychology Review, vol. 9, no. 3, (2015).
5. C. D. Ingledew, "Motivation and Change in Behavior," *Psychology of Change*, vol. 11, no. 4, (2018).
6. A. J. West et al., "CREB Function in Learning and Memory," *Nature Reviews Neuroscience*, vol. 10, no. 8, (2009)
7. P. S. T. Lee et al., "Gene Regulation by CREB," *Annual Review of Genetics*, vol. 43, (2009).
8. T. C. K. Wong et al., "Metabolic Disorders and the CREB Gene," *Journal of Clinical Endocrinology & Metabolism,* vol. 99, no. 8, (2014).
9. R. R. H. Asaka et al., "CREB and Synaptic Plasticity," *Current Opinion in Neurobiology*, vol. 14, no. 3, (2004).
10. D. B. W. Smith, "Neurodegenerative Diseases Linked to CREB Dysregulation," *Journal of Neurochemistry*, vol. 125, no. 6, (2013).

11. THE FREQUENCY OF PRAYER

1. Mitchell L. Gaynor, *The Healing Power of Sound: Recovery from Life-Threatening Illness Using Sound, Voice, and Music* (Boston: Shambhala, 2013).

ABOUT THE AUTHORS

DR. SCOTT OATSVALL is a certified LT360 holistic health coach, cellular health scientist, and bioenergetic specialist with a passion to help people get healthy at the cellular level and live a healthy lifestyle. He has coached and taught health and human performance at the high school and college levels for over twenty-three years.

In 2014 he founded Life Transformation 360 (LT360), which is a cellular health and all-natural health care company designed to educate and coach people how to achieve total health, mind, body, and spirit. In 2018 he opened up the first all-natural functional and fully integrated health center in Middle Tennessee.

Dr. Scott is the host of "The LT360 Coach's Health Show." As a renowned coach, speaker, author and radio personality, Scott has helped thousands of people by delivering a high impact message which tells people how to develop the seed of greatness in them so that they can be healthy, happy and whole while being all that they were created to be.

Dr. Scott and his wife, Gwen, live in Franklin, Tennessee, with their six children. They are also the founders of 147 Team Ministries, which serve under privileged, vulnerable, and at-risk children locally and around the world.

For more information on products, scanning, services, or speaking, visit www.LT360.com or call (615)815-0390.

JOAN HUNTER is a compassionate minister, a dynamic teacher, an accomplished author, and an anointed healing evangelist. Her focus is to train and equip believers to take the healing power of God beyond the four walls of the church and into the four corners of the earth!

Joan has ministered in countries all over the world and has made numerous television and radio appearances. She hosts a powerful and exciting show of her own called *Miracles Happen!* She has authored more than thirty-one books and has recorded teachings that will encourage you and teach you how to pray for the sick and see them recover.

Joan and her husband, Kelley, live northwest of Houston, Texas. Together, they have four daughters, four sons, three sons-in-law, and seven grandchildren. Joan is the daughter of the Happy Hunters, Charles and Frances Hunter.

Connect with Joan at JoanHunter.org or JoanHunter.ca.

John Tussey is an award-winning keyboard artist, composer, arranger, producer, music educator, and public speaker who lives in Hawaii. He is a music pioneer and forerunner who has recorded with frequencies of the Periodic Table of Elements since 2011, and is one of only two people worldwide who publicly releases this music. John also records with the Solfeggio frequencies and other frequencies downloaded to him from heaven, and uses 432 Hz tuning and 444 Hz tuning.

For more information, visit John's website at www.johntussey.com.

Made in the USA
Monee, IL
13 September 2026

62979545R20085